Advances in

Agricultural Research in India

Advances in
Agricultural Reseach in India

Y.S. Sreenivas

Oxford Book Company
Jaipur ▪ New Delhi

Published By

Oxford Book Company

Regd. Off. "Bony Residency", Gate No. 2
Opp. Tilak Public School, Vishweswariya Nagar,
Gopalpura Road, Jaipur - 302018, India
Ph: 0141 - 2761280, Telefax: 2761381
email - oxfordbookjpr@gmail.com
website - oxfordbookcompany.com

Delhi Office: 102, 1st Floor, Satyam House,
4327/3, Ansari Road, Darya Ganj,
New Delhi - 110002, India
Ph: 011-45652440

Distributed By
KEYA INTERNATIONAL PRESS LLP
E-mail: keyainternationalpressllp@gmail.com

Published in India, 2024

ISBN : 9788189473792

Price: ₹ 1925.00

Typeset by Shivangi Computers, Jaipur

Printed and bound in India by Thomson Press Ltd.,
New Delhi

Preface

The agricultural strategies adopted in the 1960s in India, which brought forth the Green Revolution, have revolutionised Indian agriculture. Pushing into the background the era of chronic and continuing food strategies, it provided a breakthrough in India's long closed circle of rural poverty and spearheaded a movement that made agriculture the backbone of India's economy. Nonetheless, the demand for food is ever increasing, and in such a scenario, it becomes pertinent to adopt newer and better methodology of agricultural practice to confront these challenges. Agricultural research thereby becomes and essential matter of concern.

This text has been intended as a manual for farmers agronomists and agricultural engineers on the progress of agricultural research in India. The aim is to not just provide an overview of the situation, but to gain an insight into the challenges and issues confronting the agricultural arena and how agricultural research procedures, techniques and breakthroughs are handling it. Serving information on the prospects, advantages, problems and the future offered by agricultural research, the book intends to foreground the issue within the contemporary context of rising food process and decreasing food quality. The critical perspectives should enhance better understanding of the matter.

Y.S. Sreenivas

Contents

1

Agricultural Research Priorities

The agricultural scenario is dominated by small land holders in the South Asia region, with a few exceptions in some areas or sectors. The importance of agriculture, though central to the economic development, is declining over time in relative terms. The prime concern of all the countries in the region was to attain self-sufficiency in food, and a number of agriculture development programmes were initiated to achieve this objective. All these countries introduced land reforms such as redistribution of surplus land, ceiling on land holdings, protection of agricultural tenancy, consolidation of land holdings, etc. to accelerate agricultural growth.

The performance, however, differed from country to country and the impact was limited due to lack of supportive systems, like input supply, credit, marketing, etc. It is now widely acknowledged that due to non-availability of these supportive systems, the agricultural growth bypassed the resource-poor farmers and the regions. For instance, owing to the differences in supportive institutions, the rice productivity in eastern India is still far below that in the north-west India.

The most important sources of growth in agriculture, particularly in India and Pakistan, are non-price factors like technology, education etc. The research investments in surface irrigation and development and dissemination of improved technologies contributed largely to the agricultural growth, ushering the Green Revolution in the region. The HYVtechnology along with a regular supply of fertilisers and water could lift the production frontier up during the 1960s and 1970s. This increase in the productivity attracted private investments in agriculture. In addition, incentives in the form of subsidised inputs and remunerative output prices also attracted private investments in agriculture.

More recently, macro economic reforms, introduced in some countries including India during 1990s, have further accelerated the agricultural growth. These reforms, on one hand, encouraged private investments (both domestic and foreign) in infrastructure and supportive system, on the other, improved the incentives in agriculture through better terms of trade, accelerating agricultural growth. Although it is rather premature to establish direct impact of these reforms at this stage, it is believed that the reforms would create conducive environment for productivity led knowledge-intensive agriculture.

PRODUCTIVITY AND AVAILABILITY OF FOODGRAINS

In spite of high population pressure and no or only a limited expansion of arable land, the countries in South Asia have made tremendous progress in terms of achieving self-sufficiency in foodgrain production. The production of cereals during the last three decades, has doubled reaching a level of 245 million tonnes in 1999. The production of pulses has varied from 12 to 15 million tonnes during the last four decades. Another remarkable achievement, albeit less discussed, is that milk production in the region has increased more than three times during this period.

Most of these gains, however, were negated by the enormous growth in population. Consequently, the annual per capita foodgrain production remained almost stagnant (around 180 kg) during 1960s to 1980s and increased moderately to 197 kg during 1990s. Similarly, in spite of almost four-fold increase in milk production, the per capita production increased from 48 kg in 1961 to 80 kg in 1999. Nevertheless, there is a marked decline in the import of foodgrains and the region is becoming self-sufficient in food production.

Another significant achievement on food security front is the stabilisation of production and prices of foodgrains in the region. It has been widely documented that year-to-year fluctuations in foodgrain production had registered a significant decline not only in favourable irrigated environment but also in rainfed regions. This had significant implications for food security of the region. In spite of floods, droughts and cyclones, there were few instances of starvation, large imports and food aids. This coupled with a better management of foodgrain stocks and the integration of domestic market, and assured availability of food. The prices of foodgrains decreased in real terms and remained more stable than the international prices.

Notwithstanding these significant achievements, the crop yields are still low in the region—yields of rice (clean) and wheat are less than 3 tonnes/ ha. The productivity of agricultural workers is also very low. The level of fertiliser consumption is moderate and barring few irrigated pockets, extent of farm mechanisation is also low. The limited area under irrigation with no scope of its further expansion and the declining per capita availability of arable land in the region call for increasing crop productivity. In the hill and mountain regions, increasing the productivity of fruits and foresting in a sustainable manner, could be very important.

TREND IN FOOD DEMAND

Two major changes have been observed in the trends in the food demand in South Asia:

— there is noticeable decline in per capita consumption of cereals, particularly coarse cereals, probably because of a decrease in the real prices of cereals and increase in real income; and
— consumption pattern has become more diversified because of increase in demand for high-value products like fruits, vegetables, milk and meat.

These changes have important implications on food and nutritional security. These countries have not only to produce additional food but also diversify food production towards products of higher nutritional value. The second concern of food security is that the demand for food would increase because of the rise in population, income of poor people and feed demand. It is estimated that the demand for foodgrains in South Asia would increase to about 360 million tonnes by 2030 AD, assuming a moderate to high rate of growth in income (3.5 to 5.5 per cent per annum).

Depending upon the growth in income, demand for milk would be in the range of 192-232 million tonnes and that for fruits, 110-138 million tonnes. An increase of a similar magnitude is expected in the demand for vegetables, meat, fish and eggs. It is important to note that for meeting such large increase in demand, yields of foodgrains would have to be increased by about 50 per cent by 2030 AD. The required increase in yields of livestock and other high value commodities would be in the range of 100-200 per cent, depending upon the rate of income growth. These targets of yield increases are quite challenging.

SUSTAINABILITY OF AGRICULTURAL SYSTEMS

The concerns relating to sustainability of agricultural systems are becoming central to the development process. These concerns have been studied and explained by a number of researchers in various ways. A widely accepted measure is the agricultural total factor productivity (TFP)—productivity of a system by taking all outputs and inputs together. It is observed that there is a deceleration in the growth of TFP in the agriculturally developed (irrigated) regions. It is also observed that a number of constraints like buildup of pests, depleting soil fertility, weeds, etc. are emerging in the irrigated production systems.

The most important concern is related to the sustainable use of natural resources. It is being felt increasingly that natural resources—land and water—are depleting fast. Land degradation due to salinity, alkalinity, water-logging, overgrasing and erosion by water and wind is widespread and unabated. Intensification of land-use, NPK imbalance, less application of organic manure, and adverse effect of pesticides on microbial activities in soil, are fast eroding the productive capacity of land. These sustainability issues need to be addressed, whilst enhancing the productivity of agricultural systems.

Table 1. Status and efficiency of groundwater-use in south Asia

Country	*Bangladesh*	*India*	*Nepal*	*Pakistan*	*Sri Lanka*
Annual water resources 1990 (km^3)	2357.0	2,085	170.0	418.3	43.2
Total withdrawals 1990 (km^3)	23.8	518	2.9	155.7	8.7
Per capita withdrawals 1990					
—Domestic (m^3)	7	18	6	26	10
—Industry (m^3)	2	24	2	26	10
—Irrigation (m^3)	211	569	143	1226	483
Irrigation effectiveness 1990 (%)	30	40	58	49	36
Percentage increase in the withdrawals in 2025 over 1990					
- With current level of irrigation effectiveness	89	67	122	134	51
- With 70% irrigation effectiveness	2	15	87	91	-4
2025 withdrawals					

(with 70% irrigation effectiveness) as percentage of annual water resources	1	29	3	71	19

In a recent study, Seckler et al. have examined the present status and future requirements of groundwater resources. It has been indicated that most of the groundwater is used for irrigation purposes, and irrigation effectiveness is less than 50 per cent in South Asia. Further, with the current level of low irrigation effectiveness, withdrawals of groundwater would increase by 67 per cent in India and 134 per cent in Pakistan by 2025 AD, which could be brought down to 15 and 91 per cent, respectively, if irrigation effectiveness is increased to 70 per cent.

With such a marked increase in irrigation effectiveness, India and Pakistan would still withdraw 29 and 71 per cent of their groundwater resources, respectively in 2025. It is important to note that these are the average figures for these two countries and the situation of groundwater-use is alarming even today in semi-arid and arid regions.

Agricultural Development Issues

The development issues identified for the region based on the foregoing discussion, are:

Efficient growth: The acceleration of agricultural growth would continue to be a pressing need of the region. It is not only essential to enhance the rate of growth but also to achieve an efficient growth. Higher growth in agriculture is desirable for food and nutritional security, higher employment and income, whereas the improved efficiency of production systems is essential for making agriculture competitive in the wake of liberalisation regime. Also, the growth should be diversified in terms of products base and widely spread regionally.

Poverty alleviation: It is now widely accepted that the growth in agriculture, led by the technological developments, made a significant impact on rural poverty alleviation. Given the level of absolute poverty and hunger in this region, the need for accelerating agricultural growth would exist for a long time. This growth needs to be equitable in terms of crops/commodities, regions and class of producers.

Sustainability: The issues relating to sustainability of agricultural systems are becoming increasingly important and visible. They primarily deal with the inter-generational equity in use of natural resources and protection of environment. It is necessary that the productivity level should be enhanced and sustained over time. At the same time, natural resources and environment should be protected for use by future generations. With the present widespread degradation of land, water, and genetic and other environmental resources, sustainability of agricultural system would be central to all developmental programmes in the region.

In addition, there could be a number of other developmental issues, such as export promotion, gender equity, system diversification, self-reliance, etc. Agricultural research is expected to contribute to these developmental objectives in South Asia. The agricultural research in fact, has a comparative advantage in contributing to the objectives of efficient growth and promoting sustainability of agricultural production system. It is because the development and dissemination of appropriate technologies not only provide a lasting contribution (unless replaced by a new technology) to these objectives, but their application on a large scale is cost neutral also. As regards the poverty alleviation, analysts believe that economic policy options are better placed to address this objective.

Today, in the 21st century, we are to deal with a knowledge-based society. Science, however, holds the key for development. For the countries in the region, it is critical to utilise the benefits of the new science and technology for the socio-economic development, particularly in alleviating rural poverty. Many of the rural poor depend on agriculture for employment and income. Accelerated agricultural growth offers a potential source of poverty reduction. Agricultural research should therefore play a central role in this task.

AGRICULTURAL RESEARCH SYSTEM

Intensity and organisation of research. The intensity of agricultural research, measured as: the number of scientists with at least masters degree or expenditure on research as percentage of AgGDP, varied considerably. India has the largest agricultural research system in the region, employing about twenty-two thousand scientists and spending a slightly less than 0.5 per cent of AgGDP on agricultural research and education. Research intensity is much low in other countries, with a spending of less than 0.3 per cent of AgGDP on agricultural research and education. This is much smaller than what is spent by the developing countries on an average (0.5 per cent) and certainly much smaller than that spent by the developed countries (2.5 per cent). Unlike in the developed countries, most of the agricultural research in this region is conducted by public research organisations.

The organisation of agricultural research is almost similar in all the South Asian countries. There are central as well as provincial research organisations, particularly in the large countries like India and Pakistan. There are institutes dealing with research as well as agricultural universities for education and research. At the centre, there is a council to plan, coordinate and conduct agricultural research, education and frontline extension (transfer and refinement of new technologies). The Indian Council of Agricultural Research (ICAR) is the largest and the oldest organisation in the region.

Major research thrusts. Over the years, the public research organisations have successfully addressed the research needs of their countries. As noted earlier, in the beginning, the main objective of the system was attainment of food self-sufficiency, which has now been expanded with the addition of such objectives as equitable growth, sustainability of production systems, diversification of product-mix, export promotion, etc. In terms of commodity coverage, focus has slowly expanded from research on crops to livestocks, horticulture, fisheries, forestry and natural resources. A similar expansion has taken place in the disciplines of agricultural sciences, and currently, the focus is on agricultural biotechnology.

Need for research prioritisation. The need for prioritisation of agricultural research arises because of three reasons:

- There is a considerable expansion in the research agenda and making allocation of research resources is difficult for fund providers. The conventional approaches for resource allocation are inadequate, and therefore, use of a formal and systematic approach of research prioritisation requiring detailed information, analysis and participation of stakeholders is warranted. The new approach objectively assesses the impacts of alternative research activities in terms of attainment of objectives.
- The research intensity is very low and therefore it is essential to use available resources judiciously for maximising research benefits. Also, research prioritisation would help improve the efficiency in the system.
- The fund facilitators can easily support research programmes if these have already been identified in a consultative, bottomup approach.

AGRO-ECOREGIONAL RESEARCH

Delineation and Characterisation

The agro-ecoregional basis of research planning is receiving increasing acceptance all over the globe, as it helps in targeting research efforts and achieves economies of scale through integration of research efforts. This approach requires identification and characterisation of various ecoregions based on agro-climatic and socio-economic factors. A number of studies have identified different agro-ecoregions (AERs) in the South Asia.

The Centres of the Consultative Group on International Agricultural Research (CGIAR) have also identified four broad regions in South Asia for identification of research priorities. We have used this information and our own evaluation to identify and characterise major AERs of South Asia. This has been done through refinements of broad zones considered by the CG centres, and integration of other exercises carried out by the NARSs. The zones so obtained were further refined with feedback from the stakeholders and experts. The identified AERs are:

— Hot Arid (HA);

— Semi-Arid (SA);

— Irrigated Sub-Humid (ISH);

— High Rainfall Humid (HRH);

— Sub-Humid to Humid Coasts (SHC); and

— Sub-Humid to Cold Arid Mountains (SCAM).

The geographical spread and cropping pattern of these AERs are shown in Map 17 and Map 2, respectively. All these AERs are fairly uniform, except the rainfed humid and mountain regions where there is some variability in climate, soil type and irrigated area. The Semi-Arid, High Rainfall Humid, and Irrigated Sub-Humid AERs are quite large, occupying 38.1, 26.4 and 19 per cent, respectively of the total net sown area in South Asia. They contribute about one-fourth each to the total value of output. It may be noted here that the High Rainfall Humid AER largely practising rice-based production systems, is of greater significance as it has enormous potential for further growth, and a large proportion of poor people live in this region.

The Irrigated Sub-Humid system practises ricewheat, cotton-wheat and sugarcane-wheat cropping systems. Both canal and tubewell irrigation are intensively used along with other modern inputs like fertilisers. Livestock is important in all the systems, but horticultural crops are widely grown in the Semi-Arid and the Coastal ecoregions. Another important characteristic is that, all ecoregions except Arid and a part of the Irrigated ecoregions, receive significant amount of precipitation which can be conserved and used for agriculture.

The estimates of poverty by agro-ecoregion are not readily available, but considering the administrative regions covered under various agro-ecoregions, it can easily be seen that most of the poor people are concentrated in the High Rainfall Humid, Semi Arid and Mountain agro-ecoregions. These systems are also characterised by low productivity and vulnerability to natural resources for degradation. These considerations have significant bearing on identification of research priorities.

PRIORITISATION OF COMMODITIES

Studies on research priority setting are generally carried out using five methods, singly or in combination. These are: congruence (weighted criteria) model, economic surplus model/benefit-cost analysis, mathematical programming, econometric models and simulation model. The scoring model can also be applied at micro-level for prioritisation of research projects. The choice of the model is guided by the level of priority setting (macro or micro) and availability of data, analytical skills and resources. We have applied the modified congruence model because of the ease of its application under a situation where time and data are the binding constraints.

In simple words, the congruence model allocates research resources in proportion to the relative value of production by region or commodity. It implicitly assumes that opportunities for research are equal across commodities, and that the research benefits are proportional to the value of output. The analysis is based on the present values and assumes constancy of relative shares. These restrictive assumptions imply that results of this exercise provide only a starting point in rationalising research resource allocation. The CGIAR and the Indian agricultural research system also applied this approach because of its simplicity, transparency and flexibility.

Prioritisation of commodities and regions involves calculation of an initial baseline matrix consisting of the value of output from different commodities in different regions. A composite baseline is then developed using the value of output (efficiency), number of poor people (equity), and arable land (sustainability) indicators, using equal weights for these three parameters.

These parameters capture extensity dimensions. Initial priority determination based on extensity parameters was modified using intensity parameters, viz. growth in AgGDP, per capita income, child malnutrition, extent of groundwater withdrawals, per capita water availability and number of scientists in the national system. The following are specific steps in arriving at the initial and final baselines, indicating relative priorities:

Modification of Initial Baseline

The initial base line does not fully consider the intensity dimensions of growth, equity and sustainability, and, therefore, appropriate intensity parameters or modifiers are used for modifying the baseline. The idea is that a higher priority should be given to that region where intensity of the problem is severe. For example, the agro-ecoregions with high groundwater exploitation should be accorded a high priority. Here, the direction of impact of modifier is positive.

On the contrary, the agro-ecoregions with low per capita income (indicating intensity of inequality) should be accorded a high priority. In this case, the direction of impact is negative. Thus, the selection of modifiers becomes highly crucial at this stage. Having selected the modifiers, the next step is to decide the weight and sign to be attached to each modifier while quantifying its impact on the initial baseline. The sign of the modifiers should be appropriately considered to target the impact of the modifier in the desired direction while modifying the initial baseline.

The following step is involved in quantifying the impact of modifiers:

$$\text{Impact of modifiers (Cij)} = [1+\{M_{ij} / \text{Max}(M_{ij})\} \times W_j]\ B_i$$

where, M_{ij} denotes the data for jth modifier for the ith agro-ecoregion, Max (M_{ij}) denotes the maximum value of jth modifier, and W_j is the weight for jth modifier.

Modifiers may have positive as well as negative impact on initial baseline. Above formula holds true for the modifiers having positive impact. In case of modifiers carrying negative sign, the direction has to be reversed. This is done by subtracting the standardised value of modifier ($M_{ij} / Max\ (M_{ij})$) from 1 and then multiplying by weight and the initial base line. The impact of each modifier is aggregated to get the total impact of all the modifiers. Using this aggregate impact, the initial baseline is modified by using the following steps to get the final baseline.

$$\text{Adjusted baseline } (Di) = B_i + \sum_{i=1\ n}^{k} Cij$$

New priority distribution or final baseline

$$(Ei) = (Di / \sum_{i=1}^{n} Di)*100$$

Priorities of Commodities

The relative emphasis on different agro-ecoregions based on the final baseline varies considerably from the relative priority ranking based on value of production (VOP) alone. The shifts in relative emphasis among different regions have to be translated in terms of priorities of commodities in every agro-ecoregions. This is achieved by adjusting the VOP of each commodity in each agro-ecoregion. The factor for adjustment (for each agro-ecoregion) is the ratio of the final baseline and VOP. The adjusted VOP is used for arriving at commodity priorities in an agro-ecoregion or aggregate priorities for all the agro-ecoregions.

Since data for all the modifiers by agro-ecoregions were not available, the research prioritisation amongst different agro-ecoregions was done using extensity parameters only. Although this is a limitation, one should not expect major changes in results, as the aggregate impact of all the modifiers (deviation between final and initial baselines) in the country-level analysis was less than 2 percentage points for a country. The parameters for prioritisation and weighting schemes were decided on the basis of the information provided by the NARSs.

The value of production was computed using international prices, with adjustments for freight charges. The freight charges were added to the international prices under importable hypothesis, and subtracted under exportable hypothesis. Transport cost within the region could not be considered because of non-availability of data. For internationally non-traded commodities, the domestic prices of larger producing countries were taken after converting them into US dollars. For this purpose, exchange rates reported by the International Monetary Fund were used. Necessary data for this exercise were taken from FAOSTAT and other published sources for the period 1997 to 1999, and the analysis is based on the triennium average.

Commodities and Agroecoregions

The modified congruence model gives priorities by commodities and agroecoregions. This priority matrix can be used to arrive at different priority dimensions, such as AER priorities (sum over commodities by AERs), commodity priorities (sum over AERs by commodity) or commodity group priorities for the region (sum over commodities and AERs). In this exercise, AER priorities, and commodity priorities within and across AER have been discussed here. For the benefit of national programmes, commodity priorities by countries have also been presented.

The 'priority score' is the share of a commodity/group or AER/country in 100 (per cent), and therefore, higher the score, higher is the priority. The national systems can use the priority matrix for allocation of resources across commodities or AERs. Fund facilitators can also use the priority matrix to track priority AER and commodity or *vice versa*. Since identification of research priorities was the major objective of this exercise, we have focussed on AER and commodity priorities.

Table 2. Agro-ecoregion priorities in South Asia

(Percentages)

Agro-ecoregion	*Value of production*	*Distribution of poor*	*Arable land distribution*	*Initial baseline*
Hot arid (HA)	2.91	2.4	7.4	4.2
Semi-arid (SA)	25.40	25.0	38.1	29.2
Irrigated sub-humid (ISH)	28.59	23.8	19.8	23.8
High rainfall humid (HRH)	26.63	38.8	25.7	30.1
Sub-humid to humid	10.36	4.7	5.7	6.9

coasts (SHC)				
Sub-humid to cold arid mountains (SCAM)	6.11	5.3	3.4	4.9
All agro-ecoregions	100	100	100	100

The AER priorities in South Asia are shown in Table 2. The ISH, SA and HRH are the three top priority AERs in South Asia. The efficiency objective can be better addressed by focusing on ISH and HRH, but for poverty alleviation, HRH and SA are more important. Sustainability issues are equally important in these AERs, although factors affecting sustainability may vary. For example, it could be depletion of groundwater and soil nutrients in the ISH, whereas soil erosion due to water may be more important for the other two regions. Amongst the three smaller AERs, the SHC and SCAM are more important from the points of view of productivity and poverty.

The priority commodity groups (among 91 commodities) in South Asia were found as cereals, livestock, horticultural crops (fruits and vegetables), cash crops and plantation crops in that order. Cereals were more important in all the AERs, but their priority score was 41 in the ISH and 51 in the HRH ecoregions. Livestock has been found important in all the AERs, but it got very high priority score in the HA (41) and SCAM, whereas fruits, cash crops and plantation crops were priority commodities for the SA, ISH and SHC systems, respectively.

These priority scores were obtained using importable hypothesis for foodgrains, cotton and sugar, as these were not regularly exported from South Asia. For commodities under regular exports, such as jute, rubber, tea, coffee, etc., exportable hypothesis was used. In the second scenario, exportable hypothesis was also considered for foodgrains, cotton and sugar. Results under both the scenarios showed only marginal differences in the priority scores.

The priority score of cereals and cash crops decreased marginally under the exportable hypothesis, while it improved for livestock. But considering substantial increase in demand for food in South Asia and its implications on food insecurity, we have subsequently discussed the results of the importable hypothesis also for these commodities. The priority scores of individual commodities, as given by the modified congruence approach, were used to classify commodities into high, medium and low priority commodities individually for each AER.

Commodities not covered in this table were of very low priority. Except for the HA, rice is a high priority commodity in all the AERs, while wheat is a high priority commodity in the HA and ISH, and of moderate priority in the SCAM and the HRH. Small ruminants, oilseeds and pulses are of high priority commodity in the HA and SA, whereas milch animals are of high priority in all the AERs, except in the SHC. All fruits, in general, are of low priority in all the AERs, except banana in the SA and SHC.

Table 3. Sensitivity analysis of commodity priorities for South Asia

Commodity group	*Base scenario*	*Priorities with VOP adjusted with growth in the demand*
Cereals	35.25	31.67
Roots & tubers	2.43	2.39
Pulses	4.45	4.31
Oilseeds	4.71	4.61
Vegetables	5.22	5.84
Fresh fruits	9.25	10.24
Dry fruits	0.35	0.38
Cash crops	10.53	9.97
Livestock	20.46	23.10
Plantation	4.82	4.56
Fish	2.54	2.92

Futuristic Considerations

The modified congruence analysis, which assumes constancy of relative shares of commodities or agro-ecoregions, can be a starting point for research prioritisation. But the results need to be adjusted for expected changes arising from unfolding of growth opportunities, research capacity and challenges of globalisation. But consideration of these changes requires additional data and detailed analysis. We have considered the growth opportunities by modification of baseline priorities with the growth in AgGDP.

A similar modification of the baseline with the number of agricultural scientists has also been attempted to capture research capability of the NARSs. However, some major changes are expected to emerge because of trade liberalisation; these could be income and price impacts, affecting food demand, and effect on trade depending upon competitive advantage. These effects are of greater consequence and hence must be incorporated in the analysis and the results should be examined for their sensitivity. However, implications of competitive advantage on agricultural research can be best captured at microlevel (research programmes and projects) research prioritisation, and therefore. Incorporation of changes in demand for commodities at the macro-level (commodity or ecoregion) is important because ensuring food security is one of the main objectives of NARSs in the region.

The demand projections for foodgrains include food as well feed demand. Expected changes in the demand are likely to affect prices and output of commodities and therefore this can be best captured by modification of the VOP. The VOP of a commodity was adjusted with the expected growth in its demand in the region. Since research and extension lag is about 8-11 years, the growth was extrapolated over a period of 10 years. This adjustment in the VOP implies that the commodities with higher expected growth in the demand should get high priority.

The adjusted VOP thus obtained along with the parameters of sustainability and equity was used for another iteration of the analysis. Cash crops and plantation crops also showed a moderate decrease in their priority scores, while other commodities showed no significant change. It may be noted here that in this sensitivity analysis we are making changes in the relevant parameter carrying one-third weight. Therefore, one should not expect drastic changes in the priority scores, unless weight are also changed significantly. It is important to mention here that these results on commodity priorities are only indicative in nature and some degree of scientific judgement need to be applied to capture other relevant external factors and opportunities (including chances of research success) in setting research priorities at the micro-level.

PRODUCTION CONSTRAINTS

Having identified the ecoregion and commodity priorities, the next logical step is to translate these commodity priorities into research programmes. This needs identification and prioritisation of production constraints (for priority commodities or production systems), and incorporation of growth opportunities and scientific feasibility. A survey of the available studies on the topic provides a fairly good understanding of the generic production constraints in the various AERs. These production constraints are further classified into three categories:

— natural resource-related constraints,

— other technical constraints, and

— socio-economic constraints.

Inadequacy of data does not permit us to analyse relative importance of these three types of constraints, but as felt by participants of the expert consultation and reported in some studies, these constraints did cause significant production losses. For example, abiotic stresses like drought and submergence, caused significant production losses of rice in the eastern India. Decreasing profits because of high capitalisation of production systems and depletion of natural resources, particularly groundwater, are serious binding constraints in the ISH ecoregion. Production environment is becoming more hostile in the Arid and Semi-Arid ecoregions and opportunities for employment and income growth are less.

Diversity of production systems, low infrastructure development and technology penetration, lack of markets, labour migration, etc. are major constraints to development of hill and mountain agriculture. Livestock, which is important to smallholders and landless labourers for generation of employment and income in all the ecoregions, is constrained by a number of factors, such as poor nutrition due to non-availability of feed and fodder, high incidences of diseases and less developed markets and other infrastructure facilities.

Production losses due to socio-economic constraints in all the ecoregions and sub-sectors of agriculture are also significant but difficult to estimate. A systematic strategy to address all these constraints successfully through harnessing scientific opportunities should guide further prioritisation of research programmes for various AERs.

GROWTH OPPORTUNITIES

Assessment of growth opportunities through application of science is a difficult task, but some judgement can be made using demand side considerations, clients' needs and scientific opportunities. There are a number of areas having tremendous growth potential and filling technology gap in these areas would help tap this potential. The value addition in agricultural products through agro-processing has not received due attention in South Asia so far.

Considering the extent of post-harvest losses, particularly in fruits, vegetables and other perishables, the scope for value addition, income supplementation, and employment generation is enormous. Their collective impact on poverty alleviation would be substantial. However, it requires a close collaboration with the private sector, large investments in infrastructure and detailed study of the market demand.

Nevertheless, rehabilitation of forests, protection of biodiversity and environment, interactions between forestry and agriculture, market development for non-timber and minor forest products, agro-forestry development, etc. were considered to be high impact areas. Another growth opportunity could be the management of rainwater in water deficit areas such as the arid and semiarid ecoregions. There is a need for further refinement and management of technologies and approaches for harvesting and use of water like watershed management. Adoption of these technologies requires group or community action and therefore, educating the community on this aspect would facilitate rapid adoption of these technologies.

Diversification towards employment and income generating activities like livestock and horticulture requires adequate technological and infrastructural support. In particular, their linkages with crop sector should be properly understood and exploited for complementarity. Advances in molecular biology and biotechnology can help in identification and utilisation of tolerance to various abiotic and biotic stresses, besides making improvements in shelf-life and quality of products.

Biotechnology can also play a significant role in organic farming. With the application of these tools it would be possible to reduce the time lag between research and technology development and the development of improved plant varieties and animal breeds. It would also increase chances of successful research. However, utilisation of these frontier sciences and knowledge-intensive technologies needs higher capital investment, inter-institutional linkages, effective regulatory mechanism and delivery system. Diversification of production systems through promoting livestock, fishery, bee keeping and horticulture, and its integration with marketing system would offer novel opportunities in the region.

RESEARCH PRIORITISATION TECHNIQUE

Currently, no formal research prioritisation technique was applied to identify the system-specific research priorities because it has enormous cost, (time and resources) and data requirements. The priorities are simply 'consensus judgements' of the expert groups. These groups had, however, used the following systematic processes and objective criteria to arrive at these priorities: *Firstly,* the root cause analysis was carried out to find the major production constraints and the emerging research issues were examined along with research gaps and opportunities. *Secondly,* the emerging issues were further subjected to their likely impact on improving the efficiency and sustainability of production systems and alleviating the food insecurity and poverty. *Thirdly,* the comparative advantage of the region and chances of research success or scientific feasibility were considered.

The fund facilitators may find these generic priority areas adequate to channel research grants, but individual organisations of the NARSs in the region may further fine tune these priorities for developing their own focussed research agenda. The conservation of natural resources (land water and germplasm) is extremely important and the priority AERs are: the Arid, Semi-arid, and Irrigated sub-humid. Research issues relating to the rice-based production systems in the HRH region assume high priority because of their likely impact on poverty alleviation. Socio-economic research issues relating to efficient organisation of production including agro-processing, sustainable use of resources, risk management, transfer of technologies and integration of markets, are extremely important for all the AERs.

— Assessment of poverty in the region is a matter of concern for all. Intensive efforts to study the poverty, its mapping and assessment of nature of interventions and investment priorities are to be made.

— The management and sustainable use of natural resources (biodiversity, land and water) is another important priority area for all the agro-ecoregions. Efforts are needed to assess and map the nature and extent of degradation of these natural resources. The study of technological and institutional interventions for sustainable use of natural resources is also important. These research areas are of 'public good' nature and therefore public research organisations at the national and international levels may have to pool their resources and jointly make efforts to address these research issues.

— The livestock, horticulture, fishery and forestry sectors, which have shown significant growth in the recent past, are yet to be developed fully. Concerted research efforts in these areas would diversify the sources of income and employment in the region, and can contribute to alleviation of poverty. It may be noted here that these sub-sectors are important in all the AERs, and therefore, a significant amount of economies of scale in research can be realised if attempted at global/ regional levels. Also, the private sector can be a useful ally in the R&D in these areas.

— Studies on commercialisation of agriculture and

integration of markets would help the countries to compete in the world market.

— A good amount of efforts are needed to study the institutional arrangements for improving farmers' access to technologies, seeds, credit, market, etc. Also, there is a need to assess appropriate institutional arrangements for reducing the impact of risk. Involvement of private sector (profit as well as non-profit) for these purposes and its linkages with public organisations need to be considered under an institutional perspective.

REFERENCES

Alston J. M., C. Chan-keng, C. Marra, P. G. Pardey and D. J. Wyatt, *A meta analysis of returns to agricultural R&D. IFPRI Research Report 13*, Washington: International Food Policy Research Institute, 2000.

BARC, *A vision for agricultural research*, Dhaka: Bangladesh Agricultural Research Council, 2001.

Desai, B.M. (ed.), *Agricultural development paradigm for the ninth plan under new economic environment*, New Delhi: Oxford & IBH Publishing Co., 1997.

Evenson, R. E., C. Pray and M. W. Rosegrant, *Agricultural research and productivity growth in India*, Research report 109, Washington,DC: IFPRI, 1998.

Fujisaka, S., L. Harrington and P. Hobbs, "Rice-wheat in South Asia: systems and long-term priorities established through diagnostic research", *Agricultural Systems* 40. 169-187, 1994.

2

Agricultural Research and Education Systems

Historically, agricultural research, extension and education in India have been in the public domain. The theory of public goods holds true for research and therefore, agricultural research was conducted in the research organisations which were administered and funded by the state. Besides this, lack of capital in private sector and low appropriability of technologies did not attract significant private investment in agricultural research in the country.

The research and extension policies and regulations also were framed in the context of public institutions. However, research environment has gone a sea change over time and therefore, funding and execution of agricultural research and extension have changed accordingly.

CONTEMPORARY INSTITUTIONAL STRUCTURE OF THE NARES

Agricultural Research and Education System

In the early stage of education system in India, agricultural science was in the domain of public funded general universities, as a part of natural sciences. With advancements in science, agriculture, mainly crop science, was separated from natural sciences, but was still taught in the general universities. Crop research to some extent was also conducted. The development of independent agricultural research and education institutions can be traced back to the late 19th century.

The process started with the pioneering efforts of Lord Mayo, the then Governor General of India, leading to the establishment of Department of Revenue, Agriculture and Commerce in the Imperial and Provincial Governments in 1871. The Department was strengthened by adding staff after the report of Famine Commission. Main functions of the Department of Agriculture, as defined in the resolution of 1881, were agricultural enquiry, improvement and famine relief. During the last decade of the 19th century, experts were recruited in the Department of Agriculture, and research and teaching in agriculture and forestry was started at few places. The foundation of the Imperial Bacteriological Laboratory was laid at Pune in 1890, to start organised livestock research. It was subsequently shifted to Mukteswar in the Kumaon hills in 1893.

Agricultural research and education got major support in the first decade of the 20th century when Lord Curzon was the Viceroy of India. The most significant milestone was the establishment of the Imperial (now Indian) Agricultural Research Institute (IARI) at Pusa in Bihar in 1905. The 'Pusa' institute suffered from a devastating earthquake in 1934 and was therefore, shifted to New Delhi, a central place, in 1936.

The development of research work at the IARI over time led to the origin of a number of research institutions. Also in 1905, six agricultural colleges were established in important provinces at Pune (Maharashtra), Kanpur (Uttar Pradesh), Sabour (Bihar), Nagpur (Maharashtra), Faisalabad (now in Pakistan) and Coimbatore (Tamil Nadu) with an annual grant of Rs 2 million from the Government of India. These colleges were adequately equipped with staff and laboratories and were charged with the responsibility of research and teaching.

Another significant development was the establishment of the Imperial (now Indian) Council of Agricultural Research (ICAR) in 1929, an autonomous body, on the recommendation of the Royal Commission on Agriculture. The ICAR was mandated to promote, guide and coordinate agricultural research in the country. With a non-lapsing fund of Rs 5 million, the ICAR was expected to supplement research activities of provinces and train scientific manpower. However, the ICAR had no administrative control over research institutions in the provinces.

The establishment of the ICAR, in a way, was empowerment of agricultural research in India. Concomitantly, a number of central commodity committees were constituted, mainly for commercial crops. These committees were semi-autonomous bodies financed by grants from the Government of India and/or by income from cesses and were expected to promote overall commodity development, including research. In fact, many committees established research stations. These committees had representation of various stakeholders like producers, trade and industry, agricultural department, etc., and Vice-President of the ICAR was ex-officio President of the committee. The funding of these committees from cesses was the first attempt to link research funding with the beneficiaries.

The commodity approach to research lacked coordination between commodities and neglected research areas applicable across commodities like soil management. The need was, therefore, recognised to initiate research on cross-commodity basis. Also, the idea of regionalisation research was getting momentum. These forces led to the establishment of Composite Regional Stations for research on cotton, oilseeds millets in 17 regions in 1956. These stations were under the administrative control of the ICAR and research progress was monitored by the regional coordination committees. The research expenditure was shared by Indian Central Cotton Committee, the Indian Central Oilseeds Committee and the ICAR.

Although the ICAR was established as a coordinating body, effective research coordination was missing because the ICAR did not have administrative control over many of the central or any of the provincial research stations. In order to provide effective coordination to commodity research, the concept of coordinated research project for improvement was introduced.

In 1957, the first All India Coordinated Project on maize was started with the technical support from Rockfeller Foundation. The project was multidisciplinary in nature pooled staff working in different regions. This was the beginning research planning on the basis of agro-climatic zones, cutting ac political boundaries. The project was extremely successful and paved way for establishment of a series of all India coordinated research projects. On the recommendation of the Agricultural Research Review T, the ICAR was reorganised in 1965 for coordinating, directing promoting agricultural research in the country.

All the commodity committees were abolished and research institutes under these committees and Central Department of Agriculture and Food gradually transferred to the ICAR. This led to centralisation of funding execution and management of agricultural research with greater autonomy and empowerment to the ICAR. A Department of Agricultural Research and Education (DARE) was created in 1973 in the central Ministry Agriculture to establish direct linkages of the ICAR with central and state governments, and international organisations.

The Director General of ICAR, a scientist, was concurrently designated as Secretary to the DARE For centre-state coordination, eight regional committees were formed. Several new research institutions under the ICAR came into existence However, major expansion under the ICAR took place on the line commodity research. Funds for these research institutes were channelled through the ICAR from the central government. Research stations under the administrative control of the state governments continued to be funded by state governments. Although a number of agricultural and veterinary colleges were functioning under the Department of Agriculture in the states, agricultural education maintained a low profile. These colleges were crippled with administrative and financial constraints. There was virtually no coordination between agricultural and veterinary colleges.

The University Education Commission felt the need for establishing rural (agricultural) universities in the states. Subsequently, the two Joint Indo-American Teams endorsed the establishment of state agricultural universities (SAUs). The SAUs were set up on land-grant pattern of the American universities. The first one was started in 1960 at Pantnagar in Uttar Pradesh. The SAUs were given autonomous status and direct funding from the state governments. These universities imparted education on all aspects of agriculture on the same residential campus and integrated teaching with research and extension.

The US Agency for International Development (USAID) and the American land-grant universities helped development of SAUs in India. Subsequently, implementation of the recommendations of the Education Commission and Review Committee on Agricultural Universities streamlined their functioning and all matters related to agricultural research in the states were transferred to the universities. The regional research capacity in the states was further strengthened by establishing the regional agricultural research stations under the National Agricultural Research Project (NATP) in 1979 with assistance from the World Bank. These research stations, in different agro-climatic zones of the states, were under the administrative control of SAUs. Addressing zonal research needs and fostering linkages between research, extension and farmers were the main responsibilities of these research stations.

Meanwhile, there has been tremendous growth in non-agricultural universities and other scientific organisations, notably, Council of Scientific and Industrial Research (CSIR), Department of Biotechnology (DBT), Defence Research and Development Organisation (DRDO) and Department of Science and Technology (DST). These organisations also continued to strengthen, directly or indirectly, agricultural research and education.

The participation of industries both in public and private sectors in agricultural research was absent until 1950s. With the adoption of new seed-fertiliser technology in the mid-sixties, there was phenomenal growth in the industrial sector for the production of inputs. However, research activities in these industries were at the margin. The entry of private sector in seed research started in the 1970s with the popularisation hybrids. The passage of new policy on seed development in ' streamed seed research in the private sector, allowing participation transnational seed companies.

Agricultural Extension System

The national agricultural extension system also evolved with establishment of the Department of Agriculture in the Imperial provincial governments. Efforts to strengthen this Department continued up to the time of Independence. Agricultural extension was one of activities of the Department and no special attention was paid accelerate transfer of technology efforts. However, some isolated attempts were made to start special rural development programmes, including improvement of agriculture. Soon it was realised sporadic and adhoc programmes might not be effective and that there was a need for sustained rural (including agricultural) development programmes.

A nationwide, multipurpose extension network backed with professionals became indispensable. Consequently, 55 Comma Development Projects were started in 1952. Each project covered villages with a village level worker for a group of 10 villages. For e-project, extension officers-technical persons in agriculture, animal husbandry, cooperation, village industries and rural engineering-w provided. The programme was based on the philosophy of integrated rural development.

In 1953, the National Extension Service Programme identical to the community development programme but with less resource intensity, was launched with a view to cover the entire country 1960/61. The programme aimed to accelerate the pace of rural development, including increased employment and production by application of scientific methods in agriculture. The programme greatly emphasised the principle of development through self-help and peoples participation. The central government largely bore the cost of programme. Front-line extension work also was initiated as agricultural research system grew in the ICAR and SAUs.

A department or directorate extension was established in the ICAR institutes and SAUs. The basic objective of these departments was to conduct extension research demonstrate latest technologies, provide feedback to scientists, a provide training support to State Department of Agriculture. Besides, the ICAR started three major front-line extension projects, viz. National Demonstration Project, Operational Research Project and Lab-to-Land Project.

Another significant development in front-line extension was the establishment of Krishi Vigyan Kendras (KVKs) and Trainers' Training Centres (TTCs) in 1974. These KVKs and TTCs were aimed to improve technical literacy of farmers including rural women on the principle of 'teaching by doing and learning by doing'. These KVKs are currently managed by the ICAR institutes, SAUs and nongovernmental organisations (NGOs) with financial support from the ICAR. The central government also launched several schemes to achieve self-sufficiency in food production.

The important programmes were: Intensive Agricultural District Programme and Intensive Agricultural Areas Programme. These programmes concentrated on the transfer of 'package of practices' and supply of critical inputs to farmers. In other words, extension strategy combined technical information with the supply of inputs. However, this strategy was discontinued with the reorganisation of the extension system under the Training and Visit (T&V) System in 1974-75. The T&V system emphasised single-purpose professional extension workers, regular training of extension personnel and transfer of technology through personal contact with farmers. This concept was further strengthened through establishing research-extension-farmer linkages under the National Agricultural Extension Project (NAEP) in 1979.

CONTEMPORARY INSTITUTIONAL STRUCTURE OF THE NARES

The national agricultural research and education system (NARS), is dominated by the public sector. Although agriculture is a state subject in the constitution of India, major components of the research system were initiated and funded by the Union Government. The NARS has three main institutional set up with different mandates. These are: ICAR institutes to cater to upstream research needs, SAUs engaged in teaching and research for respective states, and ZARSs to undertake zonal-specific research.

A structural diagram depicting these institutions and their linkages with other actors like public research organisations. The direction of research linkages is shown with arrows. The ICAR is the apex body at the centre to promote, undertake and coordinate research in all fields of agriculture in the country. The ICAR is linked with the Union Ministry of Agriculture through the DARE. The Council also coordinates directly with state governments and international organisations through the DARE. The Governing Body consisting of eminent agricultural scientists, academicians, legislators and farmers' representatives as its members, is the chief executive and policy making authority and the General Body is the supreme body of the ICAR.

By the end of the Eighth Plan, the ICAR had established a network of 45 research institutes, 10 project directorates (PDs), 30 national research centres (NRCs), 4 national bureaux (NB) and 86 all India coordinated research projects (AICRPs), etc. Four research institutes have the status of national institute and the rest are named as central institutes.

A large number of the ICAR research institutes conduct basic/strategic, and applied research in discipline-based divisional set-up. IARI, NDRI, IVRI, CIFE with the status of 'deemed university' also undertake post-graduate teaching in agriculture. The AICRPs have their research centres at the SAUs and are engaged in applied research. Some AICRPs are elevated to the status of project directorate to provide backstopping research. The NRCs conduct research on specific problems in mission mode, non-divisional set-up. Although ICAR institutes are mandated to do basic and strategic research, a good deal of applied research is also conducted due to low research intensity in the SAUs.

All the ICAR research institutions are managed by the management committee and research advisory committee. The management committee has wide representation, while research committee is a body of research professionals. Most of the ICAR institutions are organised on commodity pattern and very few are resource or discipline based. The SAUs are autonomous institutions for meeting educational and research needs of the states and these are managed by the board of management and academic council. All the states have at least one SAU. The ZARSs under the SAUs are mandated to cater to research needs of the zones. The SAUs are largely funded by state governments, but they also get regular grant from the ICAR. One Central Agricultural University with funding from the Union Government is also established for northeastern states.

Some states have established new SAUs by elevating old campus to university. Although efforts were made to establish ICAR institutions in the major producing state of the mandated commodity, there appear to be some influence of political-economic factors. For example, a large number of institutions were established in the northern and southern states- the states having larger representation in the Union Ministry of Agriculture, while western and north-eastern states were given low priority. Only recently, one agricultural university with the central assistance has been established.

A large number of non-agricultural universities, government organisations and public sector undertakings are also involved directly or indirectly in agricultural research. Some universities like Banaras Hindu University, have independent faculty for agricultural research and education, while government departments or scientific organisations like DST, DBT, CSIR, DRDO, etc. conduct or support agricultural research directly or indirectly.

The public sector industrial units are also involved in agricultural research, mainly on inputs, to some extent. Private sector undertakes research for the development of embodied technologies, viz. chemical, mechanical and biological (only hybrids). However, private sector research so far is adaptive in nature and is expected to intensify in the years to come with the adoption of favourable industrial and regulatory policies. Several private foundations, both national and international, also conduct and/or invest in agricultural research in the country.

Research Linkages and Coordination

Considering the size and multi-institutional set-up of the NARS, developing research linkages and coordination is a formidable challenge. The task is further complicated by the fact that the responsibility of agricultural research and development lies with state governments. The ICAR as an apex body, coordinates research and promotes inter-institutional research linkages. Since the ICAR supports SAUs through regular grants, it has direct participation in the management of the SAUs.

Besides, regional committees were formed in 1975 to assess the status of research, extension and education in the ICAR institutes and SAUs in the eight regions of the country. These committees also make recommendations to undertake research on immediate problems of the region. Officials from the ICAR, ICAR institutes, SAUs, state line department, NGOs, members of parliament and farmers representatives are members of these committees.

Another informal but effective link between various research institutions is the cross-nomination of members in various committees and scientific panels. As noted above, these committees and scientific panels have major say in planning and management of research. Efforts are made to ensure effective use of research resources and to avoid duplication of research efforts.

Research collaboration with the CGIAR system, NARSs and research foundations overseas, etc. is operationalised by the ICAR through the DARE. However, SAUs can also directly collaborate with these international organisations. Linkages with the national private research organisations are direct. Public research institutions extend support such as supply of germplasm and training facilities to the private sector. Also, private research companies can collaborate directly with multinational companies or private research foundations abroad under the existing regulations which recently have been liberalised to a great extent.

Agricultural Extension System

Broadly, there are four major components of the Indian extension or transfer of technology system:

- agricultural extension service with the state governments,
- extension education system of ICAR and SAD system,
- extension programme of input industries in public and private sectors and NGOs, and
- special rural development programmes of the central and state governments.

However, main responsibility of transfer of technology rests with the state governments as agriculture is a state subject. The central government also implements several schemes having transfer of technology component, through the state governments.

The main extension system comprises the Directorate of Extension in the Union Ministry of Agriculture at the centre and T&V system under the state Department of Agriculture. In the T&V system, professional extension workers work with contact farmers for the transfer of information and skill. Training support to the T&V system is provided by the ICAR/SAU system.

The Directorate of Extension plans extension activities at national level and disseminates information through mass media and publication of literature. For training of extension staff, there is a three-tier training system. At the national level for training of senior and middle level staff, an autonomous institute, namely, National Institute of Agricultural Extension Management (MANAGE) was established. There are four regional extension training institutes and several state training institutes for training of extension workers.

Several development programmes like integrated rural development programme, watershed development programme, operation flood, technology mission for crops, etc. sponsored by other government departments contain transfer of technology component. In 1994, the scope of extension was widened under the broadbased agricultural extension' in farming system approach to include all landbased activities.

The ICAR/SAU front-line extension system plays a catalytic and supportive role. It develops extension methodology, refines and transfers front-line technologies, and provides feedback to scientists. This system has three approaches. First is the special transfer of technology programmes like National Demonstration (ND), Operation Research Project (ORP), and Lab-to-land Programme (LLP). Most recent in the series is the Institute-Village Linkage Programme started in 1995/96 for technology assessment and refinement. The second approach comprises the transfer of technology and training by the KVKs and TTCs on the principle of 'learning by doing'.

There are 261 KVKs functioning in the country under the ICAR institutes, SAUs and NGOs. Front-line extension programmes (ND, ORP, LLP) are presently merged with the KVKs. Eight TTCs provide training to the KVK staff. In the third approach, ICAR/SAU system provide training to the master trainers (subject matter specialists) working in the state line department through monthly workshops.

Private input companies are involved only in the transfer of chemical (fertilisers, pesticides), mechanical and biological (hybrid seeds) technologies developed/produced by them. Public sector companies and seed corporations also undertake transfer of technology activities related to the sale of their products i.e., farm inputs. Commodity groups/boards also promote commodity specific extension activities. Many NGOs also undertake extension activities as part of their development programmes. Some NGOs are also managing ICAR-supported KVKs.

Efforts have been made to institutionalise research-extension linkages at national, regional, state and zonal levels. At the national level, under the ICAR-DAC interface joint meetings of the senior officers from the ICAR and Department of Agriculture and Cooperation (DAC) are organised twice a year to discuss critical research and development issues. At the regional level, eight regional committees were constituted to review research and development status in the ICAR institutes and SAUs located in the region. These committees represented by the senior research and extension officers, farmers representatives and NGOs meet once in two years.

The Zonal Agricultural Research and Extension Advisory Committee meetings and seasonal workshops at the zonal level facilitate close interaction between researchers, extension workers and farmers. In the T&V system, research-extension linkages are institutionalised through monthly/bimonthly workshops for the training of master trainers or subject matter specialists in state line department by the researchers.

AGRICULTURAL RESEARCH AND EXTENSION SYSTEMS

An attempt also is made to assess private investment in research and extension in the country. First, combined investment in research and education is discussed as research and education are highly complementary activities and the national agricultural research system (NARS) is pursuing both these activities jointly. Moreover, the distinction between research and education expenditures was not uniformly maintained in the government accounts, making delineation of 'research' expenditure difficult. This is followed by analysis of public investment in extension.

Government investment includes the expenditure made by the Union and all the State governments, and Union Territories. These data were compiled from various official accounts of the Union and State governments (GAG, MOF and RBI) and contain all plan and non-plan expenditures on revenue as well as on capital accounts.

Funds from the Union Government support the ICAR, the apex body charged with the responsibility of policy planning, execution and coordination of research. Besides supporting a network of ICAR institutions, a part of funds are transferred to SAUs in the form of research schemes and annual grants. The State governments support SAUs which are entrusted with the responsibility of imparting education and conducting state or location specific research. Some government funds are also used to support research in public organisations like Agro-economic Research Centres and commodity research stations outside the ICAR and SAU system. Research funding from commodity boards like tea, coffee, etc. are not included here.

Investment Intensity

The real investment (centrestate) in 1994/95 registered more than five-fold increase since the 1960s. The phases of change in the real investment correspond to organisational changes in the NARS. The low and declining central investment during the late 1960s coincides with the shift from multi-channel research funding to centralised funding to the ICAR. The establishment of the SAUs accelerated the state funding in the 1960s and the state funds contributed all the growth in total investment in the country.

Reorganisation of the ICAR in 1973 and substantial increase in the investment in the Fifth Plan (1974-78) set a sharp uptrend in the central funds. Efforts to strengthen the decentralised research capacity with the implementation of the National Agricultural Research Project and much higher allocations in the Eighth Plan have further accelerated the growth in total investment. The decade-wise growth rates indicate that the total investment, in real terms grew at the rate of 5.4 per cent since 1960s. The investment made by the states grew much faster than the central in the 1960s and 1980s, whereas the growth in the latter was remarkably high during the 1970s.

Table 1. Annual compound growth rates of government real investment in research and education (%)

Period	Centre	States	Total
1960/61 to 1969/70	-16.71	9.07	6.52
1970/71 to 1979/80	6132	-0.05	9.57
1980/81 to 1994/95	5.88	6.83	6.29
1960/61 to 1994/95	11.99	3.46	5.38

The changing emphasis on the structure of the NARS over time has changed the share of centre and state governments in the national investment. As seen from Table 2, during the 1960s and 1970s state governments' funds contributed most of the total investment. Their share rose from 80 per cent in the early sixties to 97 per cent in the early seventies. However, more than proportionate increase in the central funding raised the center's share substantially (55 per cent) in the early nineties.

Table 2. Share (%) of central and state governments in the national investment

Period (three-year average)	Centre	States
1960/61 to 62/63	20.1	79.9
1970/71 to 72/73	3.3	96.7
1980/81 to 82/83	46.9	53.1
1991/92 to 93/94	44.6	55.4

As seen from Table 3, research and education intensity, as measured by the investment as percentage of agricultural (excluding forestry) gross domestic product (AgGDP), rose from 0.23 in the early seventies to 0.39 in the early eighties and to 0.49 in the early nineties. Also, the real investment per ha of gross cropped area (GCA) increased over time, reaching Rs 20.65 in the early nineties.

Table 3. Intensity of government research and education investment: All India

Indicator	*1960-62*	*1970-72*	*1980-82*	*1992-94*
1. Investment at current prices (Rs, million)	142	409	1858	9617
2. Investment at 1981/82 prices (Rs, million)	729	1081	1887	3831
3. Ratio of investment to AgGDP (%)	0.21	0.23	0.39	0.49
4. Investment/ha of GCA (Rs)				
—at current prices	0.93	2.47	10.49	51.85
—at constant prices	4.77	6.57	10.65	20.65
5. Number of scientists	—	—	—	22249
6. Investment/scientist (000 Rs)				
—at current prices	—	—	—	432.27
—at constant prices	—	—	—	172.18
7. Area/scientists (000 ha of GCA)	—	—	—	8.34

These intensity ratios indicate that the investment in research and education is growing since 1960s, but the major jump came in the 1980s, almost doubling research intensity in the country. Another important investment indicator is the number of scientists, which is not readily available. Recent data compiled by the ICAR (ICAR, 1995/96) and Rao and Muralidhar indicate that there are about 22,249 scientists engaged in research and teaching in ICAR/SAU system. With these data annual nominal investment per scientist during the early nineties works out be Rs 0.4 million. Apart from governments, industries in public and private sectors also invest in research on seeds, fertilisers, pesticides, machinery, drugs, sugar, and food and leather processing. Adding this industrial investment to government investment, gives aggregate 'research' intensity in the country. Government 'research' expenditure was obtained by taking out education expenditure from the total. The share of 'research' is 86.9 per cent in ICAR and 53.5 per cent in SAUs.

Aggregate 'research' intensity shows that during 1992-94 (three-year average), governments contributed 80 per cent to national research investment. The contribution of public sector industries was only 5 per cent, raising the share of public investment to 85 per cent. Only 15 per cent investment came from the private sector. All together the country spends 0.42 per cent of AgGDP on research, which is quite low as compared to other countries. It is about 0.5 per cent for developing countries and 2.4 per cent for developed countries.

Efforts should be made to raise the intensity to at least a commonly described norm of 1 per cent of AgGDP. Another important indicator of research intensity is the annual research expenditure per scientist. There are 11,048 full-time equivalent (FTE) scientists in the country. Of these, 3,977 are working with the ICAR and 7,071 with the SAUs. Annual nominal research expenditure per scientist is much higher in the ICAR Rs 0.9 million as against Rs 0.4 million in the SAUs, giving an average expenditure of Rs 0.6 million per FTE scientist.

Research and Education Investment

It is important to balance the factor shares in total investment. Shortage of funds for a critical factor may seriously impair the efficiency of the system. The share of various factors, namely, salary, capital and operating expenses are not readily available, especially for SAUs. The data compiled from ICAR budget book give some idea about factor share in the total expenditure. Salary or establishment cost cornered 46 per cent of the total funds and 37 per cent went to meet operating and equipments cost in the 1980/81. The share of salary rose to 63 per cent in the midnineties, whereas the share of operating and equipments cost decreased to 23 per cent. Expenditure analysis done during the preparation of Agricultural Technology Project (NATP) also indicates substantial reduction in the share of 'operating expenses' in the ICAR expenditure. The share of 'operating expenses' is even lower in the SAUs.

Since state-wise data on government investment in research and education are available over time, and government investment dominates the national investment, state-wise analysis is done for the government investment. Growth in real investment became negative in the 1970s in some states, notably, Assam, Himachal Pradesh, Jammu & Kashmir, Madhya Pradesh and Uttar Pradesh.

All the states showed impressive investment growth in the 1980s. In the 1980s, in majority of the states the annual growth was more than 6 per cent and the growth was remarkably high in the states of Madhya Pradesh (13.3 per cent) and Tamil Nadu (12.3 per cent), whereas the growth was moderate in West Bengal (2.35 per cent). Impressive growth maintained during the 1980s can be attributed to the investment made under the NARP. However, in spite of appreciable investment growth, the investment intensity remained well below 1 per cent of AgGDP in all the states, except Himachal Pradesh.

Table 4. Composition of ICAR expenditure

Factor	*Per cent share*	
	1980/81	*1995-96*
Establishment costs	46.29	63.39
Travelling costs	1.56	0.98
Operating cost including cost of equipments	37.47	23.28

Civil works	12.54	10.71
Other costs	2.13	1.64
Total	100.00	100.00

Only five states, namely, Assam, Himachal Pradesh, Kerala, Tamil Nadu and Maharashtra have achieved an intensity which is comparable with or higher than that for the country as a whole (0.49 per cent). Large and less developed states of Madhya Pradesh, Uttar Pradesh, Bihar, Rajasthan, and Orissa have very low research and education intensity (less than 0.2 per cent). Although the investment in Punjab was increasing about 7 per cent per annum, its intensity was still low. The low intensity in Punjab may be because of specialised cropping system (rice-wheat) of the state which, to some extent, provides economy of size to research efforts.

In contrast, the establishment of four SAUs in Maharashtra, mainly on socio-political grounds, and two SAUs in Himachal Pradesh have raised their research and education intensity. The increased intensity was also reflected in terms of research and education investment per hectare. Real investment per hectare increased in 1990s over 1980s in all the states, with marked increase in the states of Jammu and Kashmir, Himachal Pradesh, Kerala, and Tamil Nadu. On the other hand, marginal increase in the expenditure gave very low intensity on per hectare basis in the states of Bihar, Madhya Pradesh, Orissa, Rajasthan, Uttar Pradesh and West Bengal.

The nominal annual expenditure ranged from Rs 145 thousand in Madhya Pradesh to Rs 545 thousand in West Bengal. Only in three states, viz. West Bengal, Kerala and Maharashtra, expenditure per scientist is close to that for the country as a whole. Interestingly, none of the states has per scientist annual expenditure equal to that in the ICAR (Rs. 0.9 million), indicating that scientist in ICAR institutes are better funded.

Regional Congruence

Another important aspect of research efforts is the congruence between actual and desired allocation of research resources across states. Normative resource allocation pattern suggested by Jha et al. was computed by considering efficiency, equity, sustainability and export as research objectives. The actual investment by states is arrived at by pooling the State and ICAR investments. Investment on ICAR institutes was added to the state in which they are located. This is a weak assumption as it implies that technologies originating from ICAR institutes have equal regional spill-over/in effects. Expenditure for seven institutes, viz. IARI, NDRI, IVRI, NAARM, CIFE, IASRI and NCAP having national mandates of strategic research, was allocated among all the states. Following three criteria were used for the allocation:

IARI, IASRI, NCAP : state's share in total gross cropped area

NDRI, IVRI : state's share in total livestock population

NAARM : state's share in the number of scientists

CIFE : state's share in total fish production

Congruence index (Cl) was computed to assess the closeness of existing investment with the near optimal one, assuming equal per unit pay offs. The index is derived as: $Cl = 1-\Sigma_i (R_i-V_i)^2$; where R_i is the actual share of ith state in the national investment, and V_i is the normative share of ith state. The value of Cl is 0.99, indicating high congruency in regional investment pattern.

On the other hand, the current share is comparatively higher than the desired in the states of Haryana, Himachal Pradesh, Kerala and Maharashtra. In contrast to general belief, the share of small states comprising north eastern states, except Assam, is not less than the optimal one. The deviations between the current and optimal shares appear to be small in percentage points, but in nominal terms, one per cent change in current investment level implies a reallocation of Rs 66 million. Therefore, an enhanced research investment in the low research intensity states would maximise research benefits in the country. It is essential to raise the investment in these states. However, there is hardly any visible attempt in the planning process to increase research investment in the low intensity states.

The share of these states in the total plan investment and ICAR grants and expenditure is comparatively lower. The state of Maharashtra, which has four SAUs claimed about 14 per cent of the total ICAR grants, indicating that states with higher research investment were also getting higher ICAR financial support. Similarly, the states of Andhra Pradesh, Haryana, Kerala and Maharashtra received higher ICAR expenditure. This strategy may not optimise research benefits.

A judicious approach would be to develop regional research capacity which can maximise overall research benefits, avoiding problems of research overlapping and 'free riding'. The ICAR can temporarily bridge the gap, but a lasting solution would be to allocate higher plan funds to those states which have low research intensity and a share lower than the desired one in the national investment. Assam, Bihar, Madhya Pradesh, Orissa, Uttar Pradesh and West Bengal are in this category, higher than the desired in the states of Haryana, Himachal Pradesh, Kerala and Maharashtra. In contrast to general belief, the share of small states comprising north eastern states, except Assam, is not less than the optimal one.

The deviations between the current and optimal shares appear to be small in percentage points, but in nominal terms, one per cent change in current investment level implies a reallocation of Rs 66 million. Therefore, an enhanced research investment in the low research intensity states would maximise research benefits in the country. It is essential to raise the investment in these states. However, there is hardly any visible attempt in the planning process to increase research investment in the low intensity states. The share of these states in the total plan investment and ICAR grants and expenditure is comparatively lower. The state of Maharashtra, which has four SAUs claimed about 14 per cent of the total ICAR grants, indicating that states with higher research investment were also getting higher ICAR financial support.

Similarly, the states of Andhra Pradesh, Haryana, Kerala and Maharashtra received higher ICAR expenditure. This strategy may not optimise research benefits. A judicious approach would be to develop regional research capacity which can maximize overall research benefits, avoiding problems of research overlapping and 'free riding'. The ICAR can temporarily bridge the gap, but a lasting solution would be to allocate higher plan funds to those states which have low research intensity and a share lower than the desired one in the national investment. Assam, Bihar, Madhya Pradesh, Orissa, Uttar Pradesh and West Bengal are in this category.

Although the allocation of research funds in the ICAR and SAUs is quite wide-spread among disciplines, some adjustments are necessary. For example, allocations to agricultural engineering can be rationalised as corporate investment (public and private) is substantial in this area.

Investment in Agricultural Extension

Extension activities are supported with funds from revenue account of the state governments, and expenditure on capital account is absent, except negligible expenditure on animal husbandry for few years in some states. The estimates of government extension investment reported here, therefore, cover only expenditure on the revenue account.

Primary responsibility of transfer of technology rests with the state governments, and ICAR and SAUs are involved only in front-line extension. This fact is clearly visible from the sources of extension investment, indicating that more than 90 per cent investment is made by the states. Most of the expenditure was channelled through the 'Department of Agriculture' of state governments. Government investment on extension has grown since 1960/61, except abrupt changes in two years. In 1966/67, the investment rose sharply because of substantial increase in the investment in Bihar, Tamil Nadu and Maharashtra. On the contrary, there was drastic fall in the investment in 1974/75, particularly in Andhra Pradesh, Bihar, Haryana, Kerala, Punjab and Tamil Nadu.

In fact, this fall is more because of changes in the accounting heads of the governments, rather than systematic efforts on the part of state governments to reduce the investment. Since 1974/75 extension investment by the governments maintained steady growth of 7 per cent. Extension intensity increased from 0.09 per cent of AgGDP in the early 1960s to 0.14 per cent in the early 1970s, which further rose marginally to 0.15 per cent in the early 1990s after slight fall in the early 1980s. Annual investment in extension by governments during the triennium 1992-94, in nominal terms, is Rs 3008 million, giving an investment of Rs 16.22 per ha.

Annual expenditure per extension worker is 26 thousand, whereas it is as low as Rs 15 thousand for the main extension system, i.e., Department of Agriculture. Even making allowances for high proportion of low qualified extension workers, there are hardly any operational funds which are essential for mobility of extension workers. Macklin estimated that the share of non-salary component in total extension expenditure in Tamil Nadu decreased from 48 per cent in 1981/82 to 4 per cent in 1990/91.

Like research, private and public sector companies also undertake extension work to promote sales of their products, however, their extension expenditure data are not available. The private companies spend slightly higher amount on publicity and on an average they spend about 4 per cent of their turnover on research and 1 per cent on publicity. We have, therefore, taken 20 per cent and 25 per cent of research expenditure of the public and private companies, respectively, as extension expenditure. This is very crude, underestimation of private expenditure on extension, as private companies also spend on marketing of inputs and maintaining of field staff.

Adding the private expenditure to the government investment gives an extension intensity of 0.2 per cent of AgGDP and Rs 21.32/ha in the country. This level of extension intensity in India is much lower than that in developing and developed countries in the early eighties. Of the total investment, 92 per cent is public investment and the rest 8 per cent is made by the private sector. Among the public funds, about 76 per cent of the national expenditure is spent by the state departments, 14 per cent by the ICAR/SAU system and 2 per cent by the public sector companies. Thus, unlike research, extension is the major responsibility of the state governments. Private sector's participation in extension will grow along with the increase in its research investment and with the increase in farmers' ability to pay for extension services.

Table 5. Agricultural extension intensity: All India

	Indicator	*Investment (Rs., million) 1992-94*	
1.	Public investment	3649	(92.3)
	Government:		
	Main extension system	3008	(76.1)
	ICAR/SAU system	558	(14.1)
	Public sector industries		
	(20% of R&D cost)	83	(2.1)
2.	Private investment		
	(25% of R&D cost)	306	(7.7)

3.	Total investment	3955	(100.0)
4.	Total investment as % of AgGDP		0.20
5.	Investment (Rs)/ha of GCA		21.32

Extension intensity

State-wise intensity is presented since 1974/75 as uniformity in accounting of investment data across states was observed since 1974/75. The growth in extension investment during the period 1974/75 to 1994/95 was negative in the high productivity states of Haryana and Punjab. These states were joined by the states of Karnataka, Kerala, Orissa and West Bengal during the period 1984/85 to 1994/95. In contrast, the states of Uttar Pradesh, Tamil Nadu, Jammu and Kashmir, Gujarat and Assam, registered impressive growth in the investment during the second period. Although the nominal expenditure production environment of other states needs higher level of extension efforts. Even in Punjab and Haryana, there is a need for intensive extension work as there would be greater role of crop and resource per ha increased in all states in the early nineties, it declined in real terms in the states of Haryana, Orissa, Punjab and West Bengal.

The proportion of AgGDP spent on extension also declined in Bihar, Haryana, Karnataka, Kerala, Orissa and Punjab, and this decline is serious because the intensity was already very low (0.06 per cent or less) in these states, except Bihar. Extension intensity in Himachal Pradesh and Tamil Nadu was more than twice of that for the country as a whole. These results, thus, underline the need for increasing extension intensity in low productivity states, particularly in Andhra Pradesh, Kerala, Karnataka, Madhya Pradesh and Orissa.

The wide variations in extension intensity across the states are rather puzzling. These variations can partly be attributed to interstate differences in the funding under the NAEP. Nevertheless, these differences need further discussion. For instance, the case of Punjab and Haryana is interesting in the sense that these states have high technology adoption levels, despite of very low extension intensity.

The plausible reason could be the dominance of irrigated production environment in these states which is conducive for rapid adoption of technology. Farmer-to-farmer spread of technology is much faster in the homogeneous, irrigated production environment. Moreover, these states have very high levels of inputs use, inviting greater attention of private input companies which also undertake transfer of technology activities. Thus, heterogenous management technologies in these states and farmer-to farmer spread of management technology is comparatively slow.

It is difficult to suggest some desired level of extension intensity, however, one can judge current extension intensity vis-a-vis the task ahead. Given current stock of technologies, there is tremendous scope for yield increase. It is estimated that excepting few states, there is untapped yield potential of 40-94 per cent in most important crops of the states. This, coupled with complexity of second generation technologies and heterogeneity of production environments warrant much more intensive extension efforts. Extension services should be strengthened by scaling up investment levels and improving the quality of extension. This is especially important for low extension intensity states like Andhra Pradesh, Kerala, Karnataka, Madhya Pradesh and Orissa.

REFERENCES

Alston, J. M.; P. G. Pardey and J. Roseboom, *Financing agricultural research: International investment patterns and policy perspectives,* paper presented at the conference of the International Association of Agricultural Economists, Sacramento, California, August, 1997.

Boyce, J. K. and R. E. Evenson, *National and International Agricultural Research and Extension Programmes,* Agricultural Development Council, New York, 1975.

Dinar, A., "Extension commercialization: How much to charge for extension services", *American Journal of Agricultural Economics*, 78(1): 1-12, 1996.

Evenson, R. E., "IARCs, aid and investment in national research extension programmes", in *Research and Productivity in Asian Agriculture*, R. E. Evenson and C. E. Pray (eds), Cornell University Press, Ithaca and London, 1991.

Farrington, J.; R. V. Sulaiman and Suresh Pal, *Improving the effectiveness of research and extension systems: An analysis of institutional and socio-economic issues*, NCAP/ODI report, New Delhi, 1997.

Guttman, J. M., "Interest groups and demand for agricultural research", *Journal of Political Economy*, 86(3): 467-84, 1978.

Huffman, W. E. and J. A. Miranowski, "An economic analysis of expenditures on agricultural experiment station research", *American Journal of Agricultural Economics* (1981): 104-118, 1981.

3

Research Resource Allocation

The Imperial Council of Agricultural Research (ICAR) was established in 1929 as a national body to co-ordinate, guide, and support this effort and other initiatives of the central government. This pattern persisted through World War II. Accelerated efforts took place in the later years, as several research institutions and various commodity committees/boards were created under the central department of agriculture with proactive support from ICAR. Agricultural extension emerged as the dominant R&D strategy in the early-Independence phase. The underlying assumption was that improved technologies were available on the shelf, and that the main task was to extend these to farmers.

The national extension service and later, intensive agricultural district/area programmes sought technological upgradation of Indian agriculture. However, the agricultural crisis persisted through the 1950s and early 1960s. A consensus developed regarding the need to strengthen technical and human resource backstopping for education, research and extension at the state level through the state agricultural university ('SAD) system. A massive set-up was created over the next decade. This was a part of the 'new agricultural strategy' enunciated in the early 1960s, which assigned central role to the generation and adaptation of new seed-fertiliser technology for cereal production and backed it with the needed incentives, institutions and investments.

The ICAR was reorganised and assumed a direct role in agricultural research and education. It initiated, funded, guided, and managed a new phase of central initiatives embracing research, education, and frontline extension. Some of these were embodied in institutes, laboratories, and universities; others were in the form of multi-disciplinary, inter-institutional coordinated research programmes called All-India Coordinated Research Projects. The increasing diversification in commodity research and the focus on natural resources and regional capacity creation. It also demonstrates that the system has been proactive and has initiated lead action to minimize research lags. For example, horticulture, livestock, and fisheries research received a boost during the 1970s and 1980s, even before the growth in these sectors accelerated.

Research on natural resources and regional gaps also gained ground in the 1970s. The International Agricultural Research Centres (IARCs) have been active partners in the national endeavour all through. It presents the current status of public R&D efforts in the country under central and state aegis as well as publicly supported grassroot level voluntary bodies. In addition, there are institutions in central departments of agriculture, CSIR, biotechnology, fertiliser and chemicals, commerce, etc. which also contribute to the national effort through dedicated research institutes or research programmes.

The private sector has been a late starter. Its entry began with liberalisation of the seed sector in mid-1980s and picked momentum in the late 1970s as incentives for private investment in R&D improved. Already the number of private players has become significant and in some commodities, private research plays an important role. This is expected to gain ground rapidly.

Table 1. Status of public agricultural research system in India, 2004-05

Institutions	*Number*
Central	
National Institutes (Deemed Universities)	4
Central /Other Institutes	43
National Bureaux	5
Project Directorates	12
National Research Centres	31
All-India Coordinated Research Projects	91
Central Agricultural University	1
Krishi Vigyan Kendras	491
Zonal Coordination Units	8
State	
State Agricultural Universities	38
Agricultural /Zonal Research Stations	343

RESEARCH RESOURCES

The research resource-mix comprises the human, material, and institutional components. Trained and specialised scientists, well-equipped labs and experimental farms, necessary technical and other support services—all are woven together in institutional and organisational structures around targeted research programmes and projects. From Independence till about mid-1990s, agricultural R&D was on expansion path and it was not difficult to raise resources. The 'green revolution' generated support at the national as well as international levels. Investment and scientific manpower grew substantially and research institutions and programmes proliferated at the central as well as state levels. We began talking about one of the largest national agricultural research systems in the world.

The fiscal crisis of the 1970s changed the investment scenario. Evidence on faltering productivity and stagnating production potential prompted a more critical stance on the part of policymakers and managers of public funds. Ideas like prioritisation, resource generation, resource-sharing, privatisation, economising, and efficiency, gained ground in policy discussions. In the wake of worsening finances, the state system became critically dependent on ICAR and external grants. The former has resorted to reallocation, economising, inter-institutional collaboration, external assistance, and restraints on recruitment.

Over most of the last 40 years, there was not much concern for resources. Consequently, there has been no official attempt to systematically compile, analyse and document research resources at either the central or state level, except as part of routine financial reporting for public expenditures. Even these are not easily accessible for states. During each Plan preparation phase, there is focus on incremental resources for fresh starts to be initiated during the plan period. The massive non-plan component is rarely scrutinised. Several special projects for research monitoring were initiated during this period but these did not yield the desired information on resources or their deployment. The credit for compilation of research investment data goes to individual researchers. It has been estimated that R&E expenditures grew in real terms from Rs 2.7 billion in early-1960s to Rs 25 billion by the end of 1970s Table 2 gives the triennium average data and growth rates.

Table 2. Public investment in agricultural R&E (at 1999 prices)

Year (TE)	*Total R & E expenditure (Rs million)*	*Share of states (per cent)*	*Expenditure as per cent of agGDP*
1971	6073 (8.6)	69.2	0.32 (4.7)
1981	8007 (1.9)	52.5	0.40 (1.3)
1991	13528 (5.2)	56.6	0.45 (1.7)
2000	20773 (4.0)	50.5	0.50 (0.6)
CAGR (1971-00)	4.4 -		1.4

Research intensity also increased significantly. Contrary to the expectation that state (local) institutions would gain more importance over time, Table 2 depict increasing centralisation — the share of states in the total R&E investment has fallen from 69 percent in 1971 to 50.5 percent in 2000. Local institutions have failed to emerge as major players and supporters of agricultural R&D. This has been a major weakness which has not been well appreciated. The central system continues to press for and obtain incremental resources. States either do not bother or lack capacity to argue their case. This dynamics can only be understood in a political economy framework. Such a study is urgently needed.

Table 3. Growth in R&E investments (at 1999 prices)

Time period	CAGR (Per cent) Centre	States
1961 - 1970	10.6	7.9
1971 - 1980	6.8	-1.3
1981 - 1990	3.9	6.2
1991 - 2000	4.9	3.3
1971 - 2000	4.9	4.0

Table 2 also supports the point made earlier regarding stressed resource environment in recent years. There has been a deceleration of growth in both absolute investments and research intensity which have remained practically stagnant. Table 3 shows that while central investments have stayed more or less on course, the deceleration has been very sharp for the states and the decadal growth rates have been very unstable.

A significant development over the past decade has been the growing importance of private sector. It now accounts for nearly 13 per cent of all agricultural R&D expenditure. This is expected to grow rapidly. The implications of this have so far been discussed in terms of complimentarity and partnership. The more substantive issues of comparative advantage and selective domains remain to be tackled. These will emerge sooner than later.

MANPOWER RESOURCES

With respect to scientific manpower resources, the data situation is worse. Sectorial reviews of the research system undertaken periodically have treated this casually and selectively. The assumption of adequacy and abundance has been so pervasive that any scrutiny has not been considered worthwhile. In fact, as part of the national drive to reduce the size of the government, there is practically a ban on fresh recruitment of agricultural scientists. Ironically, neither ICAR nor SAUs have presented a factual and analytical case contesting this position.

Though authoritative data on scientific manpower resources are not available, official pronouncements and early studies do convey the impression of adequacy. By late-1980s, more than 19 thousand agricultural scientists were estimated to be engaged in the public R&E system.

There are no numbers, but recent assessments consistently reveal a decline in scientific manpower in the state system. Even the ICAR was barely able to maintain the numbers through early-1990s, essentially because of the declining level of entry-level scientists. This has happened despite escalation in the research and education agenda and is a cause of concern. Deficiencies have cropped up on the qualitative front also.

These create inefficiencies. Functional allocation and management of financial resources; level, mix, quality, and management of human resources; bureaucratic rigidities; lack of transparency and several other areas have been identified in various reviews of ICAR and SAUs. There is consensus that these are holding back research productivity.

ALLOCATIONS

Since the beginning of organised public research, resources have been created and allocated on the basis of perceptions regarding its value to the society. The research system began with a focus on improving crop and livestock productivity. This initiative of the central government expanded through provincial departments of agriculture, and persisted as an important component in the ICAR. Colonial trading interests were identified early on and research on export and commercial crops became the focus of attention at the federal level. Since Independence, research on foodgrains production occupied the centre stage, a thrust which persisted through the 1970s Even today, food security and its qualitative dimension is a core concern.

Extension was the main R&D strategy during the 1950s. The assumption was that improved technologies were available with the central and state research establishments, what was needed was its dissemination to farmers. The national extension service was built and intensified. This did not make the expected impact and priority shifted again to research and education in the mid-1960s. The high-yielding varieties programme, all-India coordinated crop improvement projects, and state agricultural universities constituted the core R&D initiatives under the 'new strategy for agricultural development'. This paid high dividends and ushered in the green revolution. Over time, neglected crops and regions were emphasised in the research agenda.

A revamped extension programme (T&V) also followed. In research, natural resource management and high-income sectors gained importance. These are the thrust areas now. Attention has also shifted to a more science-intensive transfer of technology and 'development' component, and greater role for non-governmental voluntary organisations in execution of such grassroot-level initiatives. Resources have broadly followed these policy trends. As part of the planning exercise, resource allocation decisions are made every five years. This has been a consultative process which uses judgement of knowledgeable scientists regarding current and emerging problems, prospects and opportunities in science.

Such a subjective allocation of research resources served the system well in the past; the overarching research goal of increased productivity contributed effectively to several national objectives. A number of evaluations show high rates of return to research investments and also significant contribution to poverty reduction. The scenario has become complex now and the process of resource allocation needs to be supplemented by more information and analysis. This has been hampered by lack of quantitative data on deployment of research resources, at both ICAR and SAD levels. There has been no institutional effort and individual researchers have resorted to using scientific publications data as indirect indicators of research resource allocation. Ranjitha's work showed that over the period 1965-92, there were changes in the commodity-wise allocation profile of agricultural research in states as well as nationally. She noted a shift from traditional cereal crops towards horticultural crops, pulses, oilseeds, spices and medicinal plants.

The regional distribution of research efforts for major commodity-groups was also shown. Ad-hoc attempts have been made at the national level to establish normative allocation profiles, but no institutional initiative to integrate these in the planning process has emerged. An attempt was made under NATP, but this did not explicitly figure in the allocation plans for national resources. The Tenth Five-Year Plan exercise for agricultural research remained traditional in its approach, focusing on research gaps and incremental investment needs, based on scientists' judgement regarding these parameters as well as institutional strengths and weaknesses. There is criticism that the resource-allocation profile shows persistent biases in terms of commodities and regions, neglect of evolving market opportunities, and other critical weaknesses.

On the other hand, it has been argued that in institutional and historical terms, the public research system has been responsive to the emerging needs and stresses in both commodities and resource sectors. Research resource allocation based on more analytical articulation of priorities is the need of the day. The recent initiative on establishment of PME cells at research institutions is intended to improve the situation. Ideas like constraints analysis do help in deciding the research agenda and its prioritisation, participatory research to improve relevance and adoption. IVLP, SREP, ATMA and other initiatives contribute to this and are expected to drive R&D efforts, particularly at the decentralised levels.

JUDGEMENTS FOR RESOURCE-ALLOCATION

Another fact-finding objective of the study was to quantify the resource allocation profile of the national agricultural research system. This is crucial for research policy and planning. Quantitative information has not been available in the past and research planners have used knowledge and experience-based subjective judgements for resource-allocation decisions. It has been argued earlier that such decisions can be improved by supplementing them with quantitative data and analysis.

Activity-wise Allocation of Resources

Research, teaching (training) and extension are integrated in the job description of scientists in the ICAR-SAD system. Other institutions also incorporate outreach functions in the agenda and allocate scientific manpower resources to such activities. There are no norms regarding the proportions, but the three roles are mutually reinforcing. Individual scientists allocate their time between these activities depending on their job profile, preferences, as well as perceptions regarding relative career rewards. This is the first allocation dimension considered. Table 4 shows the average participation rate of scientists belonging to different categories of institutions in these activities.

Table 4 clearly depicts that scientists in all kinds of institutions participate significantly in all the three major activities. Though research is the dominant activity across the board, teaching and extension also involve 70-72 per cent of the scientists, on an average. As expected, a relatively larger proportion of SAD and deemed university scientists are involved in teaching. Extension appears more important in local-level institutions. ICAR has an upstream research role and this is reflected in the time-allocation profile. In private institutions, research and extension are the major R&D activities in terms of participation.

Table 4. Participation rate of scientists in different activities

(Per cent)

Institutions	*Research*	*Teaching/ Training*	*Extension*	*Adminis- tration*
ICAR97.2	48.3	57.8	68.8	
Deemed universities	99.0	89.0	56.7	66.9
Other institutes	94.8	36.6	58.1	69.3
SAUs	91.8	79.6	76.6	56.7
Other public	78.6	64.3	78.4	62.3
Total public	87.3	67.7	69.6	57.4
Private	91.0	51.2	69.5	70.8
All units	91.2	70.0	72.6	60.5

Administration has been considered as a trivial activity, but the data show that more than 60 per cent of the respondents participate in this activity. Data on average time allocation to these activities provide a more precise indication of scientists, preferences and priorities. Table 5 reveals that while scientists in all institutions accord the highest priority to research, those in ICAR and private institutions spend relatively more time. Teaching is next in importance in the SAUs and deemed universities.

In ICAR and private institutions, it is not a minor claimant of scientists' time. Public institutions, particularly ICAR and SAUs, emphasise on extension and percentage involvement of scientists is high, but it is relatively undervalued by scientists in terms of time allocation. These institutions are mainly in 'frontline' extension; the state department of agriculture has the mainstream extension role. Private institutions pay relatively more attention to extension, which is more akin to sales promotion effort in their case. The 'other' public category is dominated by grassroot level units (KVKs, NGOs) and this results in high average time allocation to extension in these institutions.

Interestingly, a significant amount of scientists' time (11 per cent) is allocated to 'administration'. In ICAR as a whole, it claims more time than training or extension. This has attracted the attention of research administrators, particularly in context of the perception that there is excess of administrative and supporting staff in public institutions. The high cost this entails becomes apparent when it is realised that more than 180 scientists can be effectively added to the ICAR-SAD scientific workforce if there is a one-percent reduction in time allocated to administration! Three factors need to be kept in view in this regard.

First, scientists include R&D-related administration in this category — arranging meetings, seminars, conferences, administering collaborative research activities, ensuring proper technical specifications and quality in research-related supplies, and a number of similar activities engage their time. This is related to the second factor. Administrative and support staff in research institutions are largely recruited and carried over on the basis of government norms for generalists. It is not yet recognised that research administration is a distinct professional activity. Saddled with such support, the scientists feel more confident when they assume the responsibility themselves. Thirdly, some experience and hands-on training in this area are desirable. Nevertheless, this load can and should be brought down significantly by professionalisation of administrative and support staff in research institutions. There are no established norms in this regard and the observed time allocation profile is attributed to individual scientist's perceptions regarding relative importance of different activities in career advancement opportunities. The evaluation and reward systems in major public research institutions remain biased in favour of research. Is this right? This needs more research.

Table 5. Time allocation by scientists to different activities

(Per cent)

Institutions	*Research*	*Teaching/ Training*	*Extension*	*Adminis- tration*
ICAR	67.6	9.3	8.8	14.3
Deemed universities	57.5	23.6	7.5	11.4
Other institutes	70.5	5.1	9.2	15.2
SAUs	42.8	34.5	13.2	9.5
Other public	33.3	22.7	32.3	11.7
Total public	46.9	27.5	14.8	10.8
Private	60.4	7.5	17.1	15.1
All units	47.5	26.6	14.8	11.0

This time-sharing means that the effective research workforce is less than the 21,869 scientists enumerated in the census. The above data allow a more realistic assessment of manpower resources for research and other activities in terms of full-time equivalents (FTEs). In real terms, the number of full-time scientists in the country comes to 10,350. The estimated number is significantly higher than that estimated for late-1980s, but about 7 per cent lower than the estimate for late-1990s.

Factors like stagnation in recruitment and high rate of attrition in the ICARSAU institutions have already been pointed out. Despite all this, the Indian agricultural research system is comparable to the largest in the world. Table 6 shows that in FTE terms also, the public sector accounts for about 95 per cent of research manpower. Because of higher time-allocation coefficient, ICAR's share rises to about 30 per cent as compared to 22 per cent in terms of absolute numbers. The shares of SAUs and 'other' public institutions go down while that of private institutions rises.

These tables show that research and teaching occupy more than threefourths of scientists' time in the ICAR-SAU system; ICAR scientists assign much more importance to research on an average, but those in the deemed universities, have a pattern similar to that of the scientists in SAUs (Table 5). Professionalisation of administrative and supporting staff will permit more research and outreach activities in the public institutions with the same workforce. This is relatively more important for the ICAR-SAU set-up.

Table 6. Distribution of FTE scientists by institutions and activities

(Number)

Institutions	*Research*	*Teaching/ Training*	*Extension*	*Adminis-tration*
ICAR	3069	420	400	650
SAUs	5810	4688	1790	1290
Other public	915	624	889	322
Total public	9794	5732	3079	2262
Private	556	69	156	139
All units	10350	5801	3235	2401

A more rigorous analysis was attempted to understand the effect of these and other factors on inter-scientist variations in time allocated for research. A regression framework was employed for this analysis which was conducted separately for the ICAR and SAD scientists. The percent time allocated to research (RESTIME) by individual scientists was used as the dependent variable. The independent variables included personal attributes as well as some institutional characteristics. The following regression equation was estimated:

Restime = f(status mandate or time, age, gender, Phd, SCI, SRCI PRSCI, RMPSCI, DISANI, DISCROP, DISNRM, DISENG, DISFISH, DISHORT, DISSOC, DISOTHR)

where, STATUS dummy = 1 for national/central institutes, 0 otherwise; Mandate dummy = 1 if mandate focuses on research, 0 otherwise; Time = percent time devoted to teaching; Age = scientist's age in years; Gender dummy = 0 for women scientists, 1 otherwise; PHD dummy = 1 if scientist holds Ph.D. degree, 0 otherwise; SCI dummy = 1 for scientist/asstt professor, 0 otherwise; SRSCI dummy =1 for senior scientist/assoc. professor, 0 otherwise; PRSCI dummy = 1 for principal scientist/professor, 0 otherwise; RMPSCI dummy = 1 for research managers, 0 otherwise; DISANI dummy = 1 for animal science discipline, 0 otherwise; DISCROP dummy =1 for crop science discipline. 0 otherwise; DISNRM dummy =1 for NRM disciplines, 0 otherwise; DISENG dummy = 1 for engineering disciplines, 0 otherwise; DISFISH dummy =1 for fishery disciplines, 0 otherwise; DISHORT dummy = 1 for horticulture disciplines, 0 otherwise; DISSOC dummy =1 for social science disciplines, 0 otherwise; DISOTHR dummy =1 for other disciplines, 0 otherwise.

It should be borne in mind that the choice of independent variables was restricted by availability of data and most of these were measured qualitatively. The regression coefficients capture the net effect of individual variables. This cannot be shown by simple tabulations. The adjusted R-square values, F-statistic, and several regression coefficients are statistically significant. The ICAR regression shows that scientists in larger (national and central) institutes spend relatively less time on research. They presumably have a more diversified activity profile. Others (directorates, bureaus, centres) are primarily research-oriented and scientists concentrate on research. Teaching competes with research time, an obvious result borne out by both ICAR and SAD regressions. Interesting results were obtained with respect to variables depicting personal attributes of scientists.

Age and gender effects were discernible only in the SAD regression. For age, the result was unexpected—older scientists spent more time on research. Women scientists were found concentrating more on research. The net effects are small but statistically significant. Conflicting results were obtained with respect to educational status. In SAUs, scientists with higher degree spent more time on research, obviously those with Master's degree were assigned more teaching and other responsibilities. In ICAR institutions, the opposite seems to hold. The regressions suggest that there are some institutional differences between ICAR and SAUs, perhaps driven by varying mandate, culture, and motivation.

Both the regressions show that with advancement in career, scientists have to spend more time on non-research activities. Research time goes down. It is, therefore, important to maintain a pyramid-like career structure. The base has shrunk over the past decade. The regressions imply that this has affected research time adversely. Manpower planning has to factor in this consideration. Finally, some disciplinary bias was captured by the regressions. Crop science and natural resource management scientists spend more time on research relative to disciplines like engineering, fisheries and social sciences. Are these due to more time-intensive research processes for some disciplines? This needs to be probed further.

The ICAR regression shows that scientists in larger institutes spend relatively less time on research. They presumably have a more diversified activity profile. Others are primarily research-oriented and scientists concentrate on research. Teaching competes with research time, an obvious result borne out by both ICAR and SAD regressions. Interesting results were obtained with respect to variables depicting personal attributes of scientists. Age and gender effects were discernible only in the SAD regression. For age, the result was unexpected—older scientists spent more time on research. Women scientists were found concentrating more on research.

The net effects are small but statistically significant. Conflicting results were obtained with respect to educational status. In SAUs, scientists with higher degree spent more time on research, obviously those with Master's degree were assigned more teaching and other responsibilities. In ICAR institutions, the opposite seems to hold. The regressions suggest that there are some institutional differences between ICAR and SAUs, perhaps driven by varying mandate, culture, and motivation.

Both the regressions show that with advancement in career, scientists have to spend more time on non-research activities. Research time goes down. It is, therefore, important to maintain a pyramid-like career structure. The base has shrunk over the past decade. The regressions imply that this has affected research time adversely. Manpower planning has to factor in this consideration. Finally, some disciplinary bias was captured by the regressions. Crop science and natural resource management scientists spend more time on research relative to disciplines like engineering, fisheries and social sciences. Are these due to more time-intensive research processes for some disciplines? This needs to be probed further.

Commodity-wise Allocation

Indian agriculture is large and diverse. Varying production possibilities, long-term adaptations, ethnic and cultural preferences and other factors have added an array of primary commodities in different production systems. Most of these have been part of traditional, low-productivity agriculture. In today's context, raising productivity across the board has become a compulsion. Accordingly, the agricultural research portfolio has become large in terms of commodities. It has been shown earlier that the commodity-focus of research could be clearly identified for nearly 80 per cent of FTE scientists. The 157 commodities identified in the census were grouped into 16 categories. The broad research resource allocation profile is provided at this level. Subsequently, more details have been provided for important groups.

The crop sector dominates the agricultural research scenario overwhelmingly. Nearly four-fifths of all scientific manpower resources are dedicated to this sector as well. Livestock research claims 15 per cent and fisheries account for the rest. This crop bias pervades across institutions, public and more distinctly, in the private sector as well. ICAR plays a very important role in supporting fisheries and livestock research in the country. Other public institutions do not accord importance to the fisheries research particularly. Private research did not accord any importance to the livestock and fisheries sectors at the turn of the century. This was surprising because these have been the high-growth sectors since 1970s It should be noted that policy impediments to private investments in R&D were still formidable. This is changing rapidly.

Another reason could be that the classification scheme adopted in this study included drugs, vaccines, hormones, etc. in the resources category and not under commodities. How do various institutions contribute to research in different sectors? 'Other' public institutions also contribute about 9 per cent. ICAR puts in 70 per cent of its resources on crops research, contributing more than a quarter (26 per cent) to the national crop research effort. With most of the downstream research located in SAUs, ICAR is mandated to focus on basic and strategic research. Its clout is stronger in livestock and, particularly, fisheries research in the country. Livestock research in SAUs had many gaps and ICAR had to step in. In fisheries research, its contribution is overwhelming.

Three important points emerge. First, considering the strategic importance of crops in food, employment and livelihoods, the crop-bias in research is well merited. Commercialisation and demand-led development are relatively recent phenomena and the allocation profile reported here still reflects the traditional biases. Secondly, the which has the mandate to guide public research, has given more emphasis to livestock and fisheries research. This bears out the dynamic nature of resources allocation process in the ICAR. It has played an important role in identifying and bridging research gaps. Finally, even though the private sector is contributing significantly in the high-growth sectors, it still is a minor player in the agricultural R&D scenario and its portfolio remains locked to selected commodities. The investment environment is changing and a rapid growth in private agricultural research and broadening of the agenda are projected.

At the turn of the century, foodgrains (cereals and pulses) constituted the most important group, claiming about 30 per cent of the commodity-focused research resources in the country. Cereal research dominated this group. Self-reliance in foodgrains continues to be a very strong undercurrent, despite attainment of self-sufficiency since mid-1980s. Horticultural research comes a close second with 28 per cent. This is a diverse group–fruits and vegetables account for half of total horticultural research. Diversification and high-value agriculture has been the major thrust since the 1970s and horticulture figures importantly in this scheme. Livestock research ranks third, followed by oilseeds, fibres, fisheries, and commercial crops. These groups account for more than 98 per cent of commodity-focused agricultural research in the country. In several cases, only 1-2 per cent of resources are allocated (fodder crops, condiments and spices, medicinal/aromatic plants, flowers/ornamentals). It should be noted that more than 8200 FTE scientists are engaged in commodity research, even 1-2 per cent of this would imply 80-160 full-time scientists!

This current profile has evolved over time. Food crops and commercial crops attracted more attention till the mid-1980s. Food self-sufficiency and foreign exchange (exports and import substitution) were the overriding goals of the agricultural sector. The growth strategy began emphasizing diversification and high-value output mix since the 1970s Horticulture, livestock, and fisheries research have gained importance. Table 7 also shows the allocation pattern at the central, state and private levels.

Table 7. Allocation of research resources across major commodity groups by institutions

(Per cent)

Commodity groups	*ICAR*	*SAUs*	*Other public*	*Total public*	*Private*	*All units*
1. Cereals	16.5	26.8	20.4	22.8	27.0	22.9
Foodgrains	21.1	35.6	24.1	29.7	27.4	29.6
3. Vegetables	3.7	7.3	2.5	5.7	13.0	5.9
4. Fruits	7.1	9.4	4.8	8.2	2.8	8.0
5. Tubers	5.7	1.4	2.1	2.9	0.8	2.8
6. Plantation crops	4.5	3.8	21.1	5.6	26.9	6.4
7. Flowers/						

Ornamentals	1.3	1.2	2.1	1.3	1.5	1.3
8. Medicinal/ Aromatic	1.0	1.3	8.4	1.9	0.7	1.8
9. Condiments/ Spices	1.6	1.7	2.1	1.7	1.6	1.7
Horticulture	24.8	26.1	43.1	27.3	47.3	27.9
10. Oilseeds	7.4	11.2	4.4	9.3	4.9	9.2
11. Fibres	7.1	5.8	9.1	6.5	13.2	6.8
12. Commercial crops	7.2	3.9	2.4	4.9	2.3	4.8
13. Fodder crops	2.4	1.5	0.4	1.7	0.0	1.6
Total crops	69.9	84.2	83.5	79.4	95.0	79.9
14. Livestock	18.3	14.3	15.2	15.7	4.8	15.3
15. Fish	11.9	1.5	1.4	4.9	0.2	4.8
All commodity groups	100.0	100.0	100.0	100.0	100.0	100.0

All institutions attach importance to foodgrains but these receive higher priority at the decentralised levels. Nearly 36 per cent of SAD-resources are allocated to this category. Political economy considerations warrant this. ICAR has moderated its thrust on foodgrains focusing on basic/ strategic research, and has switched resources to the fisheries and livestock sectors where major research gaps existed. It has been able to show greater resilience and make relatively rapid adjustments, essentially because of availability of incremental (plan) funds.

The 'other' public category is dominated by small grassroot level institutions. These emphasize cereals and other commodities with high visibility and impact potential. It has been shown later that these institutions are concentrated in a few regions and local production patterns influence their portfolios significantly. Cereals also receive greater attention in the private sector research. Hybrids of maize and millets provided the entry point for the private sector in Indian agricultural research, and these along with rice hybrids, continue to be important. Plantation crops, fibres and vegetables are the other important foci for the private sector.

Table 8 shows the relative importance of different players in R&D efforts on various commodity groups in the country. The public sector overwhelms the scene and, as expected, SAUs have the leading role in most cases. Research on plantation crops, fish, and tubers is an exception to this. The ICAR accounts for about one-third of all commodity-based research; in fisheries and tuber crops research, it has a dominating share. In a few other commodities (livestock, commercial crops, fodder crops), it has strong presence. 'Other' public institutions play important role in medicinal/aromatic plants, plantation crops, flowers/ornamentals, and fibres.

Private research, has not yet assumed a comparable importance in any major group. Inter-group allocation is most balanced in SAUs and ICAR, as indicated by low CV values. 'Other' public and private R&D institutions have large variations between groups. Small size and the need to respond to specific local and market demands perhaps necessitate this.

Table 8. Share of institutions in research by commodity groups

(Per cent)

Commodity groups	*ICAR*	*SAUs*	*Other public*	*Total public*	*Private*	*All units*
1. Cereals	23.1	64.8	7.9	95.8	4.2	100.0
2. Pulses	22.0	72.9	5.0	99.8	0.2	100.0
Foodgrains	22.8	66.7	7.3	96.7	3.3	100.0
3. Vegetables	19.9	68.5	3.8	92.2	7.8	100.0
4. Fruits	28.3	65.1	5.4	98.8	1.2	100.0
5. Tubers	64.5	27.9	6.7	99.0	1.0	100.0
6. Plantation crops	22.7	33.0	29.4	85.1	14.9	100.0
7. Flowers/ Ornamentals	30.6	51.8	13.6	96.1	3.9	100.0

8. Medicinal/ Aromatic	17.5	40.2	40.9	98.5	1.5	100.0
9. Condiments/ Spices	29.9	55.5	11.2	96.6	3.4	100.0
Horticulture	28.4	51.9	13.7	94.0	6.0	100.0
10. Oilseeds	25.7	68.1	4.3	98.1	1.9	100.0
11. Fibres	33.6	47.6	11.9	93.1	6.9	100.0
12. Commercial crops	48.2	45.6	4.5	98.3	1.7	100.0
13. Fodder crops	47.6	50.3	2.1	100.0	0.0	100.0
Total crops	28.0	58.5	9.3	95.8	4.2	100.0
14. Livestock	38.1	51.9	8.8	98.9	1.1	100.0
15. Fish7	9.9	17.4	2.5	99.8	0.2	100.0
All commodity groups	32.0	55.5	8.9	96.5	3.5	100.0
Coefficient of variation	50.0	31.8	102.3		120.2	

Are different R&D institutions in accord in terms of the priority they assign to research on different commodity groups? In Table 7, some differences were noted. Non-parametric statistical tests like Kendall's test of concordance and Spearman rank-order correlations were employed to add rigour to these comparisons. This analysis was done in two steps: first, the null hypothesis that the rankings (priority) assigned to the 15 commodity groups by the four major participants—ICAR, SAUs, 'other' public, and private, were unrelated (independent) was tested by working out Kendall coefficient of concordance (W). Second, pair-wise Spearman rank-order correlations were worked out between the priority rankings observed in these institutions.

The values of W (0.669) and statistically significant Chi-square (37.46) led to the rejection of null hypothesis, implying that there was a fair agreement in the rankings assigned to various commodity groups by different institutions. One could conclude from this that all institutions are guided by the same research objectives and have similar priorities. This appears counterintuitive. Public and private R&D are segmented. Central and decentralised institutions are also expected to pursue different agenda. The concordance coefficient (W) does not seem to capture these.

Table 9 presents the full range of Spearman rank correlations between pairs of institutions. These are more revealing. The priorities of the major public players—ICAR and SAUs – are similar. ICAR's priorities appear to be independent of those of 'other' public, and private R&D institutions. This is consistent. National compulsions drive ICAR research, and these are mediated through SAUs which are the major partners. Within the same priority framework, the two institutions focus on different parts of the research spectrum—the ICAR focusing on basic/strategic research, and SAUs emphasizing on applied research.

The other two institutions have different motivations. The state and locally-oriented institutions have greater similarity as all the correlations are statistically significant. The rank correlations are more discriminating. These clearly show that the central and state components, which claim 83 per cent of national research resources, share common goals. Then, there is another cluster of institutions which have locally-driven priorities. As one would expect, the SAUs act as a bridge between the two.

Table 9. Spearman rank correlations between ranking of different commodity groups by institutions

Institutions	*ICAR*	*SAUs*	*Other public*	*Private*
ICAR	1.00	0.691	0.332	0.382
SAUs	1.00	0.588	0.617	
Other public			1.00	0.761
Private				1.00

These numbers clearly reveal as follows :

— The crop sector accounts for nearly four-fifths of all the commodity – oriented research; in private sector, the share is even higher.

— The cereals, livestock, fruits and vegetables, oilseeds, and fibre crops account for more than two-thirds of the national R&D efforts. Their share is close to three-fourths in the state system. ICAR has a more diversified portfolio. Cereals (hybrids), plantation crops, fibre crops and vegetables claim more than four-fifths of private R&D efforts. Its commitment to non-crop R&D is as yet insignificant.

— Overall priorities of ICAR and SAUs are in agreement, though specific shares of individual commodity groups differ. SAUs, 'other' public and private institutions are driven more by local needs and markets, and have broadly similar priorities. There are pointers suggesting a differentiation of public-private domains. This will grow as private R&D grows and matures.

— Public institutions have been responsive to the growing opportunities by allocating relatively higher share of research resources to livestock and fisheries research as compared to the budding private R&D sector. ICAR stands out clearly in this regard as compared to the SAUs and 'other' public institutions.

— Foodgrains research has gradually lost relative ground across the board, thanks to higher level of food production and security. This has enabled diversification to other commodities.

— Finally, the state system's dominance in most commodity research is amply brought out by the results, as is its mediating role between the central and local institutions. These have to be borne in mind as strengthening and revamping of public agricultural research system is contemplated.

Meaningful intra-group analysis is possible only with respect to major groups–cereals, pulses, vegetables, fruits, oilseeds, fibres, livestock, and fisheries. As shown earlier (Table 7), together, these eight groups employ nearly four-fifths of the total scientific-manpower resources.

Table 10. Commodity-wise allocation to cereals group

(Per cent)

Commodities	*ICAR*	*SAUs*	*Other public*	*Private*
Rice	51.6	43.8	46.8	26.5
Wheat	24.7	22.4	27.7	4.6
Jowar6.9	8.9	5.1	10.1	
Bajra	0.6	2.7	1.7	1.5
Maize	8.9	10.1	8.3	43.0
Barley	1.5	1.7	1.4	-
Other cereals	5.8	10.4	9.0	14.3

Pulses research is more diversified and nationally, gram, pigeon pea, green gram, cowpea and lentil are emphasized. ICAR's pulse research has a sharper commodity focus as compared to that of other public institutions (Table 11). Private sector has insignificant presence in pulses research, confined only to gram.

Table 11. Commodity-wise allocation to pulses group

(Per cent)

Commodities	*ICAR*	*SAUs*	*Other public*	*Private*
Gram	25.0	30.7	32.1	100.0
Pigeonpea	23.6	18.5	7.6	-
Mung	23.1	9.0	8.6	-
Lentil	10.6	2.7	3.8	-
Cowpea	1.4	5.7	26.7	-
Other pulses	16.3	33.4	21.2	-

Horticulture has the most diverse research portfolio; it

includes not only fruits, vegetables and flowers, but also other groups. Detailed break-up has been provided only for vegetables and fruits. As expected, Tables 12 and 13 suggest higher diversification in the state and 'other' public institutions.

Table 12. Commodity-wise allocation to vegetables group

(Per cent)

Commodities	ICAR	SAUs	Other public	Private
Onion	16.4	7.6	20.5	6.3
Brinjal	2.7	13.0	9.0	18.3
Cabbage	7.2	4.3	-	3.2
Cauliflower	-	1.3	9.4	7.1
Okra	4.5	9.2	-	10.1
Green pea	-	0.2	4.4	-
Tomato	15.6	16.6	6.8	18.1
Green chillies	7.7	16.2	14.0	8.5
Carrot	4.0	0.9	-	-
Mushroom	-	1.0	27.6	-
Garlic		-	1.2	4.3
Cucurbits	9.4	7.7	-	24.6
Other vegetables	32.7	20.7	4.0	3.8

Horticulture has the most diverse research portfolio; it includes not only fruits, vegetables and flowers, but also other groups (tubers, aromatic/ medicinal, condiments/ spices, etc.). Detailed break-up has been provided only for vegetables and fruits. Private research institutions focus more on cucurbits, brinjal, tomato, okra, chillies, cauliflower, and onion. We note that potato is not included in this group. Among fruits, citrus, mango, grapes, arid / dry fruits and banana are more prominent in ICAR (Table 13). Downstream institutions emphasize on guava, apple, and litchi, in addition. Papaya and banana are important in the private sector.

Table 13. Commodity-wise allocation to fruits group

(Per cent)

Commodities	*ICAR*	*SAUs*	*Other public*	*Private*
Apple	-	4.4	6.2	-
Grapes	13.2	5.4	-	-
Guava	4.7	10.3	6.6	-
Litchi	-	0.3	7.6	-
Mango	12.1	15.2	14.3	-
Papaya	3.1	2.9	2.9	28.9
Banana	6.8	10.7	13.0	33.0
Other arid fruits	-	3.7	5.8	-
Dry fruits	7.0	8.5	0.6	-
Citrus fruits	17.4	15.3	16.2	4.7
Other fruits	35.7	23.6	26.8	33.4

In livestock, cattle, buffalo, poultry, goat and sheep account for nearly four-fifths of all research resources. Livestock products are also important. The remainder is allocated to a number of other animals and birds. Institution-wise profile shows that ICAR is the lead player in camel, equines, yak/mithun, where other institutions have insignificant research and the private sector has no research at all. In terms of livestock commodities, ICAR and SAUs appear to have the same priorities. In other institutional categories ('other' public and private), the FTE numbers are too small for constructing a meaningful profile. In fisheries research as well, only a marine-inland break-up is possible.

Table 14 Commodity-wise allocation to oilseeds group

(Per cent)

Commodities	*ICAR*	*SAUs*	*Other public*	*Private*
Soybean	25.2	14.6	17.1	21.5
Groundnut	12.1	24.9	32.4	9.1
Rapeseed/mustard/toria	26.8	20.7	29.9	15.3

Linseed	-	1.1	3.0	5.0
Sunflower	-	2.6	4.0	29.8
Castor	4.5	7.8	-	19.4
Nigerseed	-	0.1	4.7	-
Coconut	13.7	7.5	7.3	-
Oilpalm	6.4	0.3	-	-
Other oilseeds	11.5	20.3	1.6	-

The current profile, clearly indicates that the inland fisheries bias is driven mainly by ICAR, perhaps in response to the rising domestic demand and historical neglect of this sector in fisheries research in the state. What is surprising is the response of the small private R&D sector which also accords relatively high priority to the inland fisheries.

Table 15. Commodity-wise allocation to fibres group

(Per cent)

Commodities	*ICAR*	*SAUs*	*Other public*	*Private*
Cotton	75.5	72.1	4.5	95.0
Jute	9.3	10.5	1.9	0.3
Sisal	1.4	0.9	-	-
Mesta & sunhemp	-	0.3	0.7	-
Other fibres	13.9	16.1	93.0	4.7

Agricultural Research

Agricultural research is mostly mediated through production resources–genetic material, land, energy, water, agro-chemicals, and so on. This is well-recognised and more than 96 per cent of the respondent scientists identified the resource focus of their research. Many researchers, for example, combined soil and water, or power and machinery in their responses. The bias in favour of germplasm resources is a global phenomenon and arises from enormous opportunities which have opened-up in this area. More than 55 per cent of research manpower is engaged in trying to raise the productivity of natural resources (germplasm, land and water). There is enormous heterogeneity in these endowments across the country which must be investigated and exploited through research. This is reflected in the data.

Material resources agro — chemicals and power/ machinery— have contributed significantly to agricultural intensification in the country and together, these claim about one-third of all resource-focused research. Agro-chemicals have been the dominant component of this with more than 26 per cent of research resources. The socio-economic environment encompassing human, institutional, infrastructural, and information resources, is an important determinant of agricultural performance and more than 9 per cent of resources are focused on it. As a caution, it should be noted that a fair amount of socio-economic research in agriculture is conducted in institutions outside the agricultural research system and the census did not cover these. Feed and fodder resources, which are important for the livestock sector, appear to be neglected even if the effort in commodity-oriented research is factored in.

Significant deviations include a relatively large emphasis on soil/water and power/ machinery research in ICAR, agro-chemicals in SAUs, and germplasm resources in 'other' public institutions (because of greater emphasis on varietal assessment work in grassroot level institutions). These are consistent. Private sector research is sharply biased towards germplasm and power/machinery and, with agro-chemicals, these account for 88 per cent of private resource-based research. It has hardly any soilwater research.

Agro-chemicals (fertilisers, pesticides, weedicides, fungicides, drugs/ vaccines, other chemicals) rank second overall. The spill-over potential in this category is believed to be high and a few good institutions nationally could meet the requirement. Yet, state-level institutions seem to accord disproportionately high priority to this sector. This is necessitated by large variations in the local agro-climatic conditions. Considerable downstream research is needed before optimal input-use strategies are finalised. Most of the SAD-research in agro-chemicals is perhaps of this genre.

Non-material resources account for a little more than 10 per cent of research resources, most of it relates to socioeconomic factors. The ICAR-SAUs system, particularly the latter, pays more attention and the private sector does not accord any priority to this area. It has been mentioned earlier that human, social, economic, and institutional resources have not traditionally been the part of agricultural research till a few decades ago. This analysis underestimates the national socioeconomics research by excluding non-agricultural research institutions.

The Kendall coefficient of concordance indicated similarities in priorities assigned to different resources by institutions. The coefficient value (0.876) had a highly significant Chi-square value (24.53), leading to rejection of independence hypothesis. The rank correlations were worked out to provide sharper details. The value indicates that there are similarities in rankings amongst all the public institutions. The private sector has a completely different orientation.

Indian agriculture is highly diverse. Agro-climatic, social, cultural and economic differentiations have given rise to a myriad of farming and production systems. Everyone agrees that agricultural development strategies must factor in this diversity. This is even more crucial for R&D planning because technology interacts strongly with agroecological and socio-economic variables. Regionalisation is, therefore, important. All public research institutions have identified relevant regions-based on their mandate. The Planning Commission and subsequently, the ICAR demarcated 15 major zones and 126 agroclimatic zones in the country, respectively.

Overall, 58 per cent of all scientific manpower resources are deployed over six zones – 6, 8, 9, 10, 11, and 12. Zones 3, 7, and 14, which are relatively important from poverty point of view, claim a relatively small (10.6 per cent) share. The green revolution focused on the Gangetic Plains and these still claim 28 per cent of national research resources. These are areas where intensification has been and continues to be pursued. The western and eastern coast regions are also high potential areas and 20 per cent of research resources are allocated there. Rainfed, semi-arid and arid agriculture dominates in Zones 7, 8, 9, 10, 13, and 14. Nearly 39 per cent of research resources are currently deployed in these regions. Together with the hill areas, these lagging regions claim 50 per cent share of national research.

Zones which have strong SAUs have larger shares. More than 56 per cent of ICAR resources are in six zones, the last four have strong state-support also. 'Other' public institutions are concentrated in the Himalayan region, Middle-Gangetic Plains and the east and west coast regions. More than 58 per cent of their resources are in these regions. Private sector investments are small, and at this time these appear to be spread everywhere. The CV value across zones is the lowest for the private R&D. ICAR has a balancing role, as indicated by its lower CV (41 percent) as compared to that of SAUs (53 per cent). 'Other' public institutions show more uneven distribution over zones (64 per cent).

One presumes that the zonal allocation profile is related to the relative size of the zones. We have used the relative share of zones in the gross area of Figures within parentheses indicate Planning Commission Zones the country as a proxy. The correlation between the two is positive (0.38) but not statistically significant. This implies that mere area of the zone is not an adequate determinant of relative priority. Value of production (VOP) is another indicator of the size in economic terms. These data are not available. As an approximation, value of crop and livestock production was used. The correlation between research and VOP shares of zones worked out to be 0.19 and was not statistically significant. It implies that size of the zone does not affect research share. It has been indicated earlier that regions pushing yield frontiers and sustainability are high on the current priority scheme, as are low productivity regions (rainfed, semi-arid and hills).

The latter are also important from poverty point of view. State institutions contribute more than 56 per cent to zonal research on an average, and nearly 30 per cent is contributed by ICAR. The SAUs share exceeds the national average in five zones (1, 6, 8, 9, 14), in all but one of these, ICAR's share is below the national average. The correlation between the two works out to be (-) 0.94. This clearly indicates that ICAR explicitly takes local research capacity into account while allocating its resources to regions. It is expected to do so. Once again the difference in ranking accorded to zones by institutions was tested using the coefficient of concordance (W) and Spearman rank correlations. The vales of W (0.523) and Chi-square (29.28) were not able to discern differences in ranking by institutions.

USE OF RESEARCH RESOURCES

Assessing efficiency of research resource use and allocation is a tentative task at most of times. R&D is a long-term and highly uncertain process. Nevertheless, research managers and institutions are compelled to model this as they allocate limited research-resources to the competing goals and programmes. Historically, their experience and knowledge of agriculture, its critical constraints, and opportunities offered by science have stood them in good stead. Ex-post evaluations showing high rates of returns to investments in agricultural research bear testimony to the efficacy of this subjective decision process.

It has been argued that research goals have become more complex now and it has become necessary to inject quantitative and diverse information and analysis in the research resource allocation decision process. Attempts have been made to develop normative research resource allocation scenarios at the national level based on multiple criteria. Such studies have pointed out that to be really able to contribute to decision-making, these must be compared with the existing resource allocation profiles. This could not be attempted because data were not available. The census data enabled some analysis along these lines in this study. This is the first such attempt at the national level.

The simplest assessment of rationality involves comparing research resource allocation with the relative economic importance of different commodities measured in terms of value of output (VOP). Estimates of such congruity, measured as congruity index. A very high level of congruence is indicated for all commodity-oriented research. An index value of 0.96 implies that, on the whole, research managers have been responsive to the criteria of economic importance of the commodity, the primary determinant of research productivity from institutional point of view. Within groups also, allocation of resources to individual commodities appears to be in line with the relative economic importance. Condiments and spices and, to some extent, plantation crops are exceptions. The former appears under-researched and the latter receives disproportionately high attention.

Table 16. Congruity index for major commodity groups

Commodity groups	*Per cent share*		*Congruity index*
	VOP	*FTE scientists*	
Cereals (8)	25.8	23.65	0.965

Pulses (7)	3.8	6.79	0.987
Fruits (12)	8.8	8.28	0.926
Vegetables (10)	8.8	6.11	0.969
Condiments/spices (13)	2.9	1.72	0.367
Plantations (4)	2.3	6.58	0.817
Oilseeds (10)	4.7	9.45	0.920
Fibres (5)	3.4	6.98	0.909
Livestock (3)	26.7	15.8	0.938
All commodities (80)	100.0	100.0	0.958

Accord between value-based priority and that revealed by relative distribution of scientists was also assessed by working out rank correlations. These have been shown in Table 17 for each institutional category. Except for the 'other' public and private categories, the null hypothesis of independence has been rejected at 1 percent probability level. This implies similarity in priority ranking; in other words, the existing research resource allocation in the ICAR-SAU system which accounts for 83 per cent of national research resources, is in line with the economic importance of commodities. In the 'other' public and private categories, research is either locally focused or targeted to few specific commodities and not to VOP rankings.

Table 17. Estimated rank correlation coefficients: VOP and FTE shares

Particulars	*Spearman rank coefficients (r s)*
ICAR	0.737
SAUs	0.786
Other public	0.317
Total public	0.754
Private	0.453
All units	0.725

Though the above analysis indicates overall efficiency in

resource allocation, it is a native approach with limited discriminatory power. A number of other objectives like sustainability, equity, trade and valueaddition, diversification, etc. have also become important. In the next step, a simple scoring model was adopted to generate a normative allocation profile. Growth, equity, sustainability, and value-addition/trade were the goals considered in this exercise; these were assigned weightages of 0.40, 0.10, 0.25, and 0.25, respectively.

The available database would easily permit evaluation of alternative weighting schemes. The final baseline (FBL) figures obtained from this exercise indicate the normative shares of important commodity-groups based on multiple criteria. In most cases, the difference between VOP share (Table 16) and modified values (FBL) are marginal. A comparison of results of this study with earlier studies showed a similar pattern in allocation of resources. But, this study emphasizes for more importance to be given to the horticulture and livestock sector. The last two columns of Table 18 are relevant for rationality assessment. A ratio (FBL/FTE) greater than unity implies the need for flow-in of resources at the cost of those where the ratio is smaller. The last column shows the needed shifts in the existing level of research resources. These shifts are on the margin. The following adjustments are indicated by this analysis.

Table 18. Optimal allocation profile and adjustment coefficients

Commodity groups	*Optimum shares*	*FBL/FTE (FBL)*	*Shift in existing resources (%)*
Cereals	0.2440	1.031	+ 3.19
Pulses	0.0401	0.591	- 40.92
Fruits	0.0923	1.115	+ 11.48
Vegetables	0.0893	1.461	+ 46.13
Tubers	0.0119	0.408	- 59.22
Condiments/ spices	0.0282	1.644	+ 64.49
Medicinal/ aromatic	0.0065	0.350	- 65.01

Plantations	0.0197	0.299	- 70.02
Commercial crops	0.0461	0.938	- 6.19
Oilseeds	0.0512	0.541	- 45.84
Fibres	0.0292	0.418	- 58.14
Livestock	0.2914	1.843	+ 84.31
Fish	0.0499	1.016	+ 1.66
All commodities	1.0000		0.00

Particulars	*Commodities*
Augmentation of resources	Cereals, Vegetables, Condiments/ spices, Fibres, Livestock
At the cost of	Pulses, Tubers, Medicinal/aromatic, Plantation crops, Oilseeds, Commercial crops

Some of these adjustments imply small percentage changes (cereals, fish), but in other cases, relatively large shifts are indicated. Some of these appear to be counter-intuitive–the suggestion to withdraw resources from oilseeds, commercial crops, pulses, for example. There are other factors like foreign-exchange compulsions, nutrition, projected scenarios, etc. which this analysis has not considered but decision-makers must reckon these. Even more important is the limitation that this static exercise does not factor in past adjustments, which go on constantly. The intention of this exercise was to make the direction of trade-off explicit. In deciding the magnitudes, policy planners have to consider these aspects also.

Table 19. Reallocation of research resources by regions

Regions	*Percent VOP# (crop and livestock)*	*FTE scientists (per cent)*	*Shift in existing resources (per cent)*
1. Western Himalayas	3.7	7.1	- 47.80
2. Eastern Himalayas	3.4	5.3	- 35.04
3. Lower-Gangetic Plains	7.5	3.4	+ 124.94

4.	Middle-Gangetic Plains	4.6	6.4	- 28.86
5.	Upper-Gangetic Plains	16.3	6.3	+ 157.98
6.	Trans-Gangetic Plains	15.4	12.2	+ 26.07
7.	Eastern Plateau and Hills	9.9	3.8	+ 159.67
8.	Central Plateau and Hills	0.3	8.8	- 96.43
9.	Western Plateau and Hills	0.4	7.7	- 94.87
10.	Southern Plateau and Hills	13.4	9.9	+ 35.64
11.	East Coast Plains and Hills	10.3	8.9	+ 15.81
12.	West Coast Plains and Ghat	4.5	11.1	- 60.06
13.	Gujarat Plains and Hills	8.0	5.4	+ 47.28
14.	Western Dry Region	2.4	3.7	- 35.54

A similar framework was employed to look at the rationality of regional allocations. This exercise was confined to VOP comparisons only because other data to construct FBL could not be collected at the regional level. Table 19 shows the VOP and the existing as well as readjusted FTE shares for 14 agro-climatic zones. In this case too, high congruity between VOP and FTE shares (index = 0.962) was indicated. However, the Spearman rank-order correlation was very low (0.20) and the null hypothesis of independence could not be rejected.

The VOP/FTE ratios suggest that Zones 3, 5, 6, 7, 10, 11, and 13 need greater attention in terms of allocation of resources in order to create a more rational resource-allocation profile. The rationalisation exercise is simplistic and three things need to be noted.

— First, a major weakness of this analysis is that it is static, it compares the actual and optimum shares in 2001-02. This has introduced bias because resource-allocation decisions factor in past decisions as well as future needs and opportunities, neither of which have been explicitly considered in this exercise. Nevertheless, in today's constrained environment such pointers would be helpful to the decision-makers. These will also assist in planning of incremental (plan) resources.

— Second, in both commodity and regional allocations, adjustments in the existing allocation of research resources have been indicated. It is emphasized that the direction of these adjustments are more important than the absolute magnitudes.

— Finally, further research in this area is urgently needed. More elaborate and sophisticated approaches have been proposed and these should be employed.

REFERENCES

Alston, J.M., *et al.*, "A Meta Analysis of Rates of Return to Agricultural R&D: Expede Herculern", *Research Report* 113. IFPRI, Washington, D.C, 2000.

Birthal, P. S., P. K. Joshi and Anjani Kumar, "Assessment of Research Priorities for Livestock Sector in India", Policy Paper 15, *National Centre for Agricultural Economics and Policy Research,* New Delhi, 2002.

Chandrashekhara, K. and K.N. Ganeshaiah, "AICRP programmes: In the need to rebirth", *Current Science,* 85 (2), 25 July, 2003.

De Datta, S. K., *Principles and Practices of Rice Production,* John Wiley and Sons, pp. 547-593, 1981.

4

Agricultural Research Policies

During the 1950's and 60's national and international concerns over food shortages, particularly in Asia, led to large investments in public agricultural research organisations. The unambiguous agenda was to raise agricultural productivity as a means of increasing the aggregate food supply. This was seen as a way of reducing hunger and the poverty associated with it. With this agenda in mind, centralised public sector scientific research institutes were created to solve the generic problem of increasing the biological potential of important food crops.

The institutional set up contained international agricultural research centres and, at the national level, sets of commodity and or disciplinary based research institutes. The task of transferring technology packages to farmers was given to the institutionally separate extension system. The task to be achieved was conceptually quite simple and all the actors in the system held this same clarity of purpose. This approach resulted in the development and spread of input responsive, high yielding cereal varieties and the consequent "green revolution" phenomenon.

NEW AGENDAS

Poverty: The green revolution, despite its success in increasing food production, demonstrated the difficulty of using advances in agricultural productivity to address complex social phenomena such as poverty. It brought to attention the fact that the poor did not always have the resources to benefit directly from new productivity enhancing technology. Moreover when the poor had land-based resources, research often had difficulty responding to their specific technological needs, usually in the less favourable production environments. For the landless and urban poor the need for better entitlements to food (through employment for example) restrained the benefits that would have otherwise arisen through cheaper and more abundant food.

Environment: Concerns were also emerging as to the environmental consequences of an intensive agricultural development strategy reliant on chemical inputs and heavy consumption of water. Wider environmental concerns were highlighted by a number of major international conferences and initiatives during the 1980's. As a consequence the environment is now a mainstream agenda for agriculture and development policy generally.

Stakeholder participation: There was also a growing realisation that the hierarchical institutional arrangements typical of most centralised, public agricultural research organisations make it difficult to achieve a client focus in research. Farmer participatory research arose as a way of addressing this issue. Sustained advocacy during the 1990's placed the participatory "paradigm" in the mainstream. Public research organisations in many countries have tried to accommodate this new agenda with varying levels of enthusiasm and success. However the institutional context of many public research organisations has restricted the development of truly participatory and client focused working practices.

Public vs. Private: In the last decade economic liberalisation along with shifts in globally held perceptions concerning the role of the State in society have emerged as a major new policy agenda. This has made institutional concerns of fundamental importance, focusing attention on the efficiency and proper role of the public sector in areas such as agricultural research. Private sector agricultural research has also grown as a result of the opportunities that economic and trade liberalisation are now presenting for private investments in agro-industries such as seed production, horticultural exports and so forth. Private research has also been encouraged by improved intellectual property protection regimes and technical advances associated with biotechnology.

Balancing agendas: The result of this ever expanding range of new policy agendas is that the "one size fits all" type of agricultural research organisation found in many countries no longer seems appropriate. Often these new agendas actually conflict with traditional internally driven policies and beliefs of the public research sector. Examples are the need to develop capacity in frontier areas of science, while also supporting adaptive research for traditional and subsistence sectors; the need to support, but not to compete with, the private sector; the need to support competitiveness in global markets such as export horticulture, but not to displace small-scale producers. All these need to be achieved without losing sight of either the old agendas of increasing food production nor the new agendas of poverty and sustainability.

The new agendas, along with the contraction of public sector funding for research in recent years, has focused attention on the need for reform of national agricultural research organisations around the world. The parallels with the Indian situation are clear.

Policy Response

A common policy response to these new agendas has been the initiation of impact assessment and research priority setting exercises. This has often been used by both national and international agricultural research organisations to achieve efficient allocation of scare funds and justify their position in terms of returns to investment. However these approaches fail to quantify impact in terms of, for instance, poverty. Another response by national agricultural research organisations has been the introduction of organisational and management (O&M) reforms. This generally includes a package of measures addressing decentralisation, autonomy, accountability, cost-recovery, transparency, and efficiency.

Institutional Innovations

Although not necessarily an explicit policy response, new and often experimental institutional arrangements are emerging in many countries as a way of addressing changing agendas and responding to emerging opportunities. These types of institutional innovations are often linked to the need to access R&D and technology more effectively and in many instances have been the result of initiatives led by private organisations both the for profit and non-profit sectors. Innovations often blur traditional institutional distinctions between research and production, usually as a way of making R&D more client responsive.

The specific features of these institutional innovations are often a response to the particular institutional set-up of a country and may therefore only be successful in this national context. Frequently these developments have significantly changed the institutional roles of different actors, with private organisations sometimes taking on roles previously performed by government.

In countries where the value of these institutional innovations has been recognised, the policy response has been to seek opportunities for public/private sector partners in research and allied activities. A related approach has been to use competitive research funds to encourage a diversity of research partners from both public and private sectors. This has often been accompanied by the separation of responsibility in the national agricultural research organisation for agricultural research policy and research implementation.

PROGRESS IN INDIA

Over the last five years the Indian Council of Agricultural Research (ICAR), being aware of the need to respond to both tightening finances, a changing research and policy environment, and a need to rationalise existing research infrastructure and human resources, implemented a series of O&M reforms. These have focused on improving efficiency and accountability; forging linkages with new partners; mobilising resources; and strengthening capacity in frontier areas of science. Priority setting exercises in major agro-ecosystems, crops and commodities have also been initiated. Reforms have not yet been applied to the State Agricultural Universities, the other major component of Indian public sector agricultural research system.

Despite some imaginative initiatives, the impact of the reforms to date has been less than expected. While there may have been some, impact on research practice, administrative and financial procedures have not witnessed any major change. Policies now exist whereby ICAR can provide consultancy and contractual services. However the extent of public/private collaborative research programmes resulting from this has not been as extensive as the potential implied by the strength of ICAR's human and scientific capital.

While there seems to be widespread agreement that there is a strong institutional dimension to this problem, the key challenge is to find ways to more effectively understand and address the issue. This is where contemporary policy analysis of the innovation process in other sectors has something to offer.

One approach a number of countries and international organisations are adopting in science and technology policy analysis in other sectors is the National Systems of Innovation (NSI) framework. The NSI provided a conceptual framework to study innovation performance in systems terms, where flows of knowledge between actors and institutions in the process are critical to innovative performance. The approach is useful to agricultural research policy at a time when changing agendas and an evolving research environment mean that while institutional change is inevitable, no consensus exists on the shape of the new system.

The contribution of NSI is that it provides a way of analysing institutional roles and relationships, and the way these change over time, and conceptualises this in terms of an innovation system. This type of analysis is more inclusive than the narrower notion of a research system. The distinction being that the latter is a system of predominantly public sector organisations engaged in producing technical innovations. In contrast an innovation system encompasses all the elements of the system or network of private and public sector institutions whose interactions produce, diffuse and use economically useful knowledge. In contrast to the research system, innovation systems are viewed as producing both technological and institutional innovations. The latter creating the evolutionary dynamics that often creates the new institutional forms that allow the creation and utilisation of new technology. This process of institutional learning is a central feature of successful innovation systems.

While the effectiveness of an innovation system usually requires strong horizontal linkages between institutional nodes, the political and social structures in many developing countries tend to encourage central control and top-down linkages, particularly in public sector organisations. The institutional innovations generated by private organisations are an implicit recognition of the shortcomings of these structures and of the benefits that can be derived from improving connectivity to other nodes and the need to change institutional capabilities to suit emerging circumstances.

INDIAN AGRICULTURAL INNOVATION SYSTEM

Recent case studies in the Indian seed and horticultural sector suggest that patterns of partnerships are starting to emerge with many of the features that the NSI recognises as important in the creation of a successful innovation system. However much of the initiative that is stimulating these developments has come from private organisations. Currently there are limitations to the extent to which—despite policy changes—the public sector can operationally contribute technology and allied R&D capability to these emerging partnerships.

Partially this relates to the administrative delays involved in contracting public sector scientist and their institutions. It also reflects the narrow research focused professional mandate of the scientists and their lack of exposure to commercial applications. Introducing policies that permit these types of contract research arrangements needs to be supported by institutional changes that allow them to work efficiently in practice. Linkages are also weak between public sector institutes, with the result that mutually useful pieces of knowledge are often locked-up in different organisations.

While scientists are often interested in engaging in partnerships with the private sector, public sector institutional hierarchies tend to engender a professional approach whereby scientifically robust recommendations are devised that can then be transferred to others. Extension organisations' continued obsession with feedback, training and demonstration and its inability to perform more valuable roles such as facilitating linkages have been also restraining the performance of the innovation system. This tends to stand in the way of developing iterative working practice that would assist in developing client-focused research. There is wide recognition by many scientists of these types of institutional constraint. However mechanisms for institutional learning and the generation of institutional innovations to address these problems are not apparent.

Agriculture the world over has changed from a resource-based to a science-based system. Technical change has accounted for half to more than four-fifth of agricultural output growth since the sixties in developed countries. In India too, as land frontiers closed down rapidly, the contribution of technical change to output growth rose from about one-fourth in the fifties to more than half since then. This trend must be fostered and accelerated. This must accompany conservation of natural resources and maintenance of ecological integrity of the agricultural system.

Technology-led growth requires investments in technology generation (agricultural" research), extension, education, physical and institutional infrastructure. Empirical studies show that agricultural research has been the primary source of total factor productivity growth in India Public investment in agricultural research is, therefore, crucial. A large and diverse country like India cannot depend on borrowing technologies from other countries. Agricultural production technologies are generally less transferable than others. Moreover, without adequate indigenous research capacity, benefits from international transfer are also minimal. Secondly Respite the current enthusiasm for privatisation, private sector cannot meet these needs fully for several reasons. Research output is often a public good.

Market failures are common and appropriability of benefits is restrictied, thus limiting incentives for private initiative. Also private sector is mainly driven by short-term profit considerations, longer term or sustainability concerns are rarely accorded priority. Therefore, even in highly developed countries, strong public research systems have to be supported. Need for this kind of support to agriculture is recognised under WTO provisions also.

CONVENTIONAL INVESTMENT CRITERIA

Use of conventional investment criteria to determine appropirate research investment levels is fraught with many conceptual and operational difficulties.

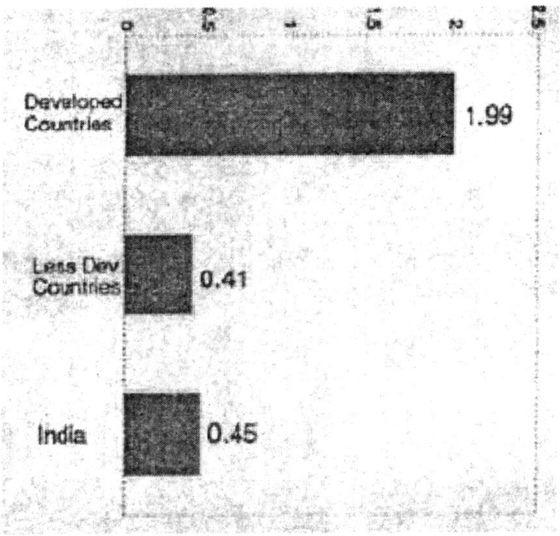

Fig. 1 agricultural research investment as percent of agricultural GDP

Rates of return studies suggest substantial underinvestment across the board. These typically show 40+ percent returns even in countries where research investment intensities are five times higher as compared to developing countries.

A norm of 2 percent of agricultural GDP for agricultural research has been suggested

— World bank, 1981, Agricultural Research: Sector Policy Paper, Washington, D.C.

— Technology Policy Statement 1993.

For India, Fig. 1 indicates that less than 0.5 percent of agricultural GDP is spent on research. At Independence, this was less than 0.1 percent, rising to 0.12 percent in 1960, and about 0.2 percent towards the end of sixties. After a spurt through the seventies, it has hovered around 0.45-0.50 percent since early eighties. This implies that research investments have barely kept pace with growth of agriculture; there has been no special accent on jacking up research investment intensity over the last 15 years.

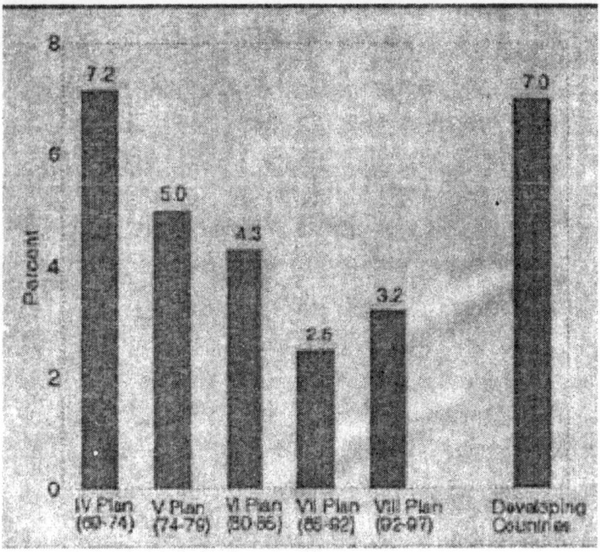

Fig. 2 Share of agricultural research and education in public spending on agriculture

It has been estimated that in developing countries as a group, research investments account for about 6-8 percent of total public spendings on agriculture. Fig. 2 illustrates the situation with regard to allocation of plan funds in India since the Fourth Five Year Plan.

The figure shows quite clearly that except during the Eighth Plan which sought to initiate a correction, agricultural research has not been accorded due priority. Since the centre accounts for nearly 60 percent of total public spending on agricultural research, signals on low priority filter down to states also. Analysis of trends in developed countries which spend close to 2 percent or more showed that between 1965 and 1985, there was positive growth in real expenditures per scientist, implying distinct improvement in research support to scientists.

In 130 developing countries, on the other hand, there was negative growth in real expenditures per scientist, suggestive of horizontal rather than vertical expansion of the research system.

Accordingly, there has been a deterioration in availability of operating funds per scientist. This has been the case for India as well. Over the Ninth and subsequent plans, these trends must be corrected and a target level of 7-8 percent of total agricultural outlay for agricultural research and education must be established. With concomitant growth in private sector spendings, this will help align total spendings on agricultural research with the needs of the agricultural sector. An aggressive R & D investment strategy is absolutely vital for agricultural growth in an internationally competitive environment.

PRODUCTIVE ARE RESEARCH INVESTMENTS

There is massive evidence from all over the world, from developed as well as developing countries, indicating that agricultural research is one of the most rewarding investment options.

A number of evaluation studies have been conducted for India. These estimates are generally based on the best available methodology and robust statistical analysis. Nevertheless, such evaluations are inherently complex and the database on research investments in the country is weak. Even if one were to discount the rates of returns estimates heavily, these still remain quite high in comparison with other investment opportunities. Indian experience is thus quite consistent with the consensus on high productivity of agricultural research.

RREFERENCES

Evenson, R.E. and D. Jha, "The Contribution of Agricultural Research System to Agricultural Production in India", *Indian J. Agri. Econ.* 28 (212-230), 1973.

Kumar, P., C.C. Maji and R.K. Patel, "Returns on Investment in Research and Extension: A study on Indo-Swiss Cattle Improvement Project, Kerala", *Indian J Agric. Econ.* 32 (3): 207-216, 1977.

Singh, A.J. and P. Kaur, "Returns to Investment on Agriculture and Extension in India and Punjab State", *Agric. Situ. India XLIV* (10): 721-26, 1992.

5

Funding and Organisation of Agricultural Research

Indian agriculture is highly diversified both in terms of production environments and activities. Small farmers (less than 2 ha) constitute about 80% of total farm holdings and occupy 40% of agricultural land. Despite a rapid increase in livestock production, the crop sector still contributes threefourths to the total value of agricultural output. Agricultural growth registered a sharp jump in the late 1960s and 1970s as a result of the widespread adoption of the new seed-fertiliser-based Green Revolution technology for rice and wheat in irrigated areas. This growth spread to rainfed areas from the 1970s with the adoption of hybrid seeds of maize, sorghum, pearl millet and cotton, although the impacts were less widespread and many areas with harsh conditions continue to experience low and unstable production.

Average crop yields have increased by an average of 1.6 percent annually over the past three decades as a result of a marked increase in irrigated area, use of modern inputs especially fertiliser, and increased cropping intensity. Yield growth and increased cropping intensity resulted in impressive growth in agricultural production, despite virtually no increase in cultivated area. Since 1980, these trends have been echoed in the livestock sector, which has grown even faster at 5 percent annually, due to rapid growth in milk, poultry and fish production.

Overall India's agricultural achievements are impressive, with increased per capita food production and accumulating food stocks. Despites these successes, India faces a major unfinished agenda with respect to agricultural productivity growth.

— First, success in reducing poverty and malnutrition, most of which are located in rural areas, is a continuing challenge. India needs to not only improve the availability of food but also generate income and employment opportunities for the poor in order to provide them the means to access food.

— Second, accelerated economic growth and rapid urbanisation are driving demand for high value commodities, particularly livestock and horticultural products, that requires that future agricultural growth become much more diversified.

— Third, sustainable management and use of natural resources is a growing challenge with depletion of groundwater, agrochemical pollution, and land degradation by water-logging, salinity, soil erosion, and deterioration of soil fertility.

— Fourth, public investment in agriculture in real terms has shown a persistent decline, while subsidies for agriculture have increased over time, despite the new economic policies.

The decline in public investment has serious implications for agricultural growth and poverty reduction. Fan et al. found that investment in agricultural research provides a high marginal return relative to other investments in terms of both growth and poverty reduction, and this return may now be higher in rainfed areas. Careful targeting of public investment, both in terms of sub-sectorial and regional priorities, and efficient utilisation of existing infrastructure, especially irrigation, is essential for achieving growth of 4% per annum contemplated in the current national agricultural policy. However, high levels of subsidies compete with funds available for needed public investment, including investment in agricultural research.

Table 1. Trends in input use, yields and production in Indian agriculture

Indicator	*1961*	*1971*	*1981*	*1991*	*2001*
Average size of holding (ha)	2.69	2.30	1.84	1.57	Na
Net cropped area (M ha)	133.20	140.30	140.00	143.00	142.20
Gross cropped area (M ha)	152.80	165.80	172.60	185.70	186.60
Gross irrigated area as					
% of gross cropped area	18.30	23.00	28.80	33.60	38.30
Fertiliser nutrient use (kg/ha)	1.90	13.10	31.80	67.40	90.00
Food grain production (M t)	82.00	108.40	129.60	176.40	203.40
Milk production (M t)	20.00	22.00	31.60	53.90	70.80
Fish production (M t)	1.20	1.80	2.40	3.80	5.40
Egg production (billion, number)	2.80	6.20	10.10	21.10	28.50
Share in the total value of production					
Crops (%)	82.40	84.40	81.40	74.70	75.70
Livestock (%)	17.60	15.60	18.60	25.30	24.30
Share of agriculture in					
Total exports (%)	44.30	36.80	35.50	22.50	26.60
Total imports (%)	36.40	37.00	18.30	11.30	11.50
Crop yields (t/ha)c					
Rice	1.01	1.11	1.29	1.74	1.94
Wheat	0.85	1.32	1.71	2.33	2.70
Coarse cereals	0.71	0.85	1.03	0.91	1.05

The current national agricultural policy envisages that market forces will guide future agricultural growth through domestic market reforms, an increasing role for the private sector, and removal of price distortions. The policy of interventions in food grain markets to stabilise prices will continue, but these interventions will be made more effective and efficient. These reforms, coupled with a focus on value addition and commercialisation, improved product quality and strengthening comparative advantage, are essential for successful transition to a knowledge-based and competitive agricultural sector. The role of the agricultural research system will be central in these processes.

HISTORICAL AND INSTITUTIONAL DEVELOPMENT

The first organised attempt to promote agricultural development, including R&E, in India began with the establishment in the last quarter of the nineteenth century of the Department of Revenue, Agriculture and Commerce in the Imperial and Provincial governments, together with a bacteriological laboratory and five veterinary colleges. These were followed around 1905 by the establishment of the Imperial Agricultural Research Institute (IARI) and six agricultural colleges in 1905. A major milestone in the history of Indian agricultural R&E system was the establishment of the Imperial (now Indian) Council of Agricultural Research (ICAR) in 1929 as a semi-autonomous body to promote, guide and coordinate agricultural research in the country. At about the same time, a number of central commodity committees were also constituted for the development of commercial crops (cotton, lac, jute, sugarcane, coconut, tobacco, oilseeds, arecanut, spices, and cashewnut).

These semi-autonomous committees, financed by government grants and revenues from a levy or cess on the output of each commodity, set up research stations for each commercial crop. Initially, the commodity committees served the interests of the imperial government (by providing revenue and ensuring raw material for industry), but later they focused on national development objectives, including research. Participation on the commodity committees was broadened to include producers and representatives of trade and industry.

An important institutional innovation in the post-independence period was the establishment of the All India Coordinated Research Projects (AICRP) initiated in 1957 under ICAR to promote multidisciplinary and multiinstitutional research. The success of the first of these for maize led to establishment of a large number of AICRPs covering all major commodities and the concept also spread to non-commodity research. In 1965, ICAR was mandated to coordinate, direct and promote agricultural research in the country, by bringing under its control all research stations previously controlled by the commodity committees and the various government departments. Subsequently, the Department of Agricultural Research and Education (DARE) was created in the central Ministry of Agriculture to facilitate linkages of ICAR with the central and state governments and with foreign research organisations.

On the recommendation of two joint Indo-American review teams, state agricultural universities (SAUs) were established on the land-grant pattern of the US. The first SAD was opened at Pantnagar in the state of Uttar Pradesh in 1960. These SAUs were autonomous in nature and funded by the government of the respective state. The SAUs integrated education with research and to some extent, frontline extension, although mainstream extension remained in the state departments of agriculture.

A number of international agencies played important roles in the development of the public agricultural R&E system in India. Notable among these were the Rockefeller Foundation for providing support to AICRPs, and the US Agency for International Development which played an active role in the establishment of the SAUs and training of faculty through partnerships with US land grant universities. The World Bank has provided considerable resources to agricultural research from 1980. The initial phase of this support emphasized the development of research infrastructure and human resources while recent support has focused on strategic research areas, priority research themes, and institutional reforms.

Current Structure of the Public Research System

Currently, the public agricultural R&E system comprises ICAR and its various institutes, and the SAUs, and their various campuses and regional institutes. At the center, ICAR funds and manages a vast research network of institutes consisting of:

- — national institutes for basic and strategic research and post-graduate education,
- — central research institutes for commodity-specific research;
- — national bureaux for conservation and exchange of germplasm and soil survey work; and
- — national research centers (NRCs) for conducting applied commodity-specific strategic research in 'mission mode'.

In addition, ICAR manages a large number of AICRPs (the coordinate programs) which draw scientists from both ICAR institutions and the SAUs. Most AICRPs centres are located on SAD campuses under the administrative control of the respective SAD. However, for the most important AICRPs (e.g., rice, wheat, maize, cattle, oilseeds, water, cropping system, and biological control of pests), ICAR has established special project directorates comprising a team of multidisciplinary scientists with their own research infrastructure under ICAR administrative control.

In 2002, ICAR had 5 national institutes, 44 central research institutions, 5 national bureaux, 10 project directorates, 33 NRCs and 82 AICRPs. In addition, ICAR established 261 *Krishi Vigyan Kendras* or KVKs at the district level that are responsible for transfer of new technologies and training of farmers. Some of these KVKs are managed by SAUs and non-governmental organisations (NGOs). In addition, there are eight training centres for 'training of trainers' in specified areas/sectors such as livestock, horticulture, fisheries, and home science. At the state level, there are now 34 agricultural universities in the country with agricultural faculties (i.e., agriculture, veterinary, engineering, and home science). Depending on the nature of the state's agriculture, SAUs may also have faculties of horticulture, fisheries or forestry and some SAUs are exclusively for animal sciences. In addition, there is one central agricultural university under ICAR to cater to the needs of small states in north-eastern India. SAUs also have zonal research stations to address research problems for each agroclimatic zone.

Besides the traditional national agricultural research system (NARS)—that is, the ICAR/SAD system—there are non-agricultural universities and organisations supporting and/or conducting agricultural research directly or indirectly. For example, the Department of Biotechnology (DBT), Department of Science and Technology (DST) and Department of Scientific and Industrial Research (DSIR) all under the central Ministry of Science and Technology support and conduct agricultural research at their institutes and sometimes fund research in the ICAR/SAD system.

Private-sector Development

Initially a few private companies dealing with agricultural inputs (e.g., pesticides, fertilisers, and machinery) invested modestly in product development, although there was little effort to establish in-house research capacity. The situation changed in the 1970s with the growing availability of trained scientists, rapid expansion of markets for agricultural inputs and processed foods, and liberalised policies to support private sector development in general.

The private sector now supplies half the certified seed, half the fertiliser, and most of the pesticide and farm machinery sales. Private investment in research currently focuses on hybrid seed, biotechnology, pesticides, fertiliser, machinery, animal health, poultry, and food processing.

The government has provided strong incentives in the form of tax exemption on research expenditures and venture capital, and liberal policies on import of research equipment to encourage participation of the private sector in research. The most significant development has occurred in the seed sector after implementation of a new Seed Policy in 1988 which allowed import of seed materials, as well as majority ownership of seed companies by foreign companies. A number of foreign seed companies entered the market, and several local seed companies have established considerable research capacity. Some local companies collaborate with overseas companies for access to proprietary tools and technologies. Private hybrids now account for a significant proportion of the market for sorghum, maize, and cotton and companies with some foreign ownership account for about one-third of this market. Developments in biotechnology have further strengthened these trends.

To provide additional stimulus to private research, India has recently approved a bill for The Protection of Plant Varieties and Farmers Rights Act to provide intellectual property protection to breeders. At the same time, it gives special emphasis to farmers' rights to save, exchange, and sell seed. India has also amended the Patent Act to make it compatible with WTO agreements. The Patent (Second Amendment) Act 1999 grants provisional product patents that should stimulate research in agricultural chemicals and the animal health sectors.

Finally, participation of private non-profit organisations in agricultural research has also increased overtime. There are now a few private foundations, as well as NGOs actively engaged in agricultural research. In particular, the MS Swaminathan Research Foundation and Mahyco Research Foundation have developed considerable research capacity with a national presence and are working in close collaboration with the ICAR/SAD system.

Contemporary Developments

The ICAR/SAD system has reached a stage where it needs to consolidate past gains through modernisation of research infrastructure, development of human capital, innovations in research management, and stronger linkages with clients. The system is responding to these challenges, albeit to varying degrees and speed.

Ecoregional research initiatives

Although the Green Revolution technologies were rapidly adopted in large areas, further gains in irrigated areas, as well as in rainfed areas that have enjoyed less benefits, require more location-specific research to adapt technologies to site and seasonal conditions. The organisation of the Indian NARS through ICAR institutes with a national or regional mandate with a strong commodity and disciplinary orientation, and SAUs based on political boundaries, has constrained the ability to respond to this challenge. Accordingly, an ecoregional approach to planning and organising agricultural research was introduced from 1978 in order to better target research efforts, integrate research across disciplines, and locate appropriate sites for research programs. Under the National Agricultural Research Project (NARP) implemented with World Bank funding, the entire country was divided into 126 agro-climatic zones consisting of several districts.

In each of the zones, a research station was established under the specific SAD to carry out applied and adaptive research relevant to the zone. An advisory committee with wide representation of farmers, NGOs, and the state department of agriculture, was created to link scientists more closely with farmers and other stakeholders, and research programs were developed through a 'bottom up' participatory approach. These zonal research stations also actively provided technical support to the KVKs and state extension departments.

The ecoregional approach was further developed under the National Agricultural Technology Project, again implemented with financial support from the World Bank. Under NATP, the country is divided into five ecoregions (arid, coastal, hill and mountain, irrigated and rainfed), which are further delineated into fourteen production systems. Research programs for each of the production systems are identified in a participatory mode and implemented using a multi-institutional and multidisciplinary systems approach. These research programs are intended to complement the AICRPs and the zonal research stations, by promoting a systems approach to planning and implementing research.

Biotechnology

Over the past decade or so, revolutionary advances in biotechnology are transforming the way agricultural research is organised and funded. To meet this challenge, the Department of Biotechnology (DBT) was created in 1986 in the Ministry of Science and Technology to support research, and human resources and infrastructure development in biotechnology related to agriculture, health care, environment and industry. DBT has established six autonomous institutions for conducting biotechnology research. It also funds biotechnology research in other institutions including ICAR institutes and SAUs, through special projects and grants, and through its competitive grants program.

In addition, ICAR has developed capacity in biotechnology research in several of its research institutes, as well as created new entities exclusively for biotechnology research. These initiatives have allowed India to develop considerable capacity in this frontier area of science, although much of it is outside the ICAR/SAD system. At least 10 research institutes have capacity in genetic engineering.

The private sector is also responding to developments in biotechnology with up to 45 companies active in agricultural biotechnology research broadly defined, for a market that was estimated to be worth US$75 million (Rs 2,663 million) in 1997. The companies include both foreign and domestic companies although all of the latter with a significant biotechnology program have developed joint ventures with global companies. At least three foreign companies have major biotechnology research facilities in India, one with a team of 34 scientists.

Given that several products are now moving into field testing and commercial release, the government is currently focusing on establishing a framework to regulate biotechnology research and the testing and release of genetically modified organisms (GMOs). The Review Committee on Genetic Manipulation (RCGM) of DBT is responsible for monitoring of biotechnology research, safety-related aspects, and import and export of GMOs. The Genetic Engineering Approval Committee of the Ministry of Environment and Forest assesses GMOs for environmental safety and approves them for wider scale testing and commercial release. India has allowed field experiments of GMOs, and approved commercial cultivation of transgenic cotton in 2002.

FUNDING OF RESEARCH

The amount of research funding and the mechanisms for fund allocation are powerful instruments of research policy in India as elsewhere. Most funds for agricultural research in India are allocated through block grants, but funding through competitive grants is now gaining acceptance, especially for operating and equipment costs.

Allocating Public Funding

Most public funding to agricultural R&E in India is in the form of block grants to ICAR and the SAUs determined according to Five Year Plans. At the beginning of each Plan, the Planning Commission constitutes a working group to agree on broad agricultural R&E priorities and to assess financial requirements for their implementation. Recommendations of the working group are discussed in several consultations between DARE and the Planning Commission. Based on the outcome of these deliberations DARE develops its Five Year Plan, and plan outlays are communicated by the Planning Commission on approval of the Ministry of Finance. This is followed by development of Five Year Plans for each ICAR institute.

Depending upon the level of proposed outlays, these plans are evaluated by committees composed of directors of the institutes, senior research managers from ICAR, and representatives of the Planning Commission, Ministry of Finance and other departments. The approved outlays are the basis for funding of each institute during the Plan period, and funds received are labelled 'Plan funds'. The on-going activities of the previous Plan are financed under 'non-Plan funding', which primarily pays salaries and other fixed costs.

A similar procedure is followed for state funding, except that state allocations are first done by the Planning Commission as part of total Plan allocations to a state. Expenditures on R&E (plan and non-plan) are then approved by the respective state governments. This process implies that resource allocation decisions are made through informed opinion and collective wisdom regarding research priorities that address developmental objectives. There is direct involvement of institutions in the allocation decisions, and input from other stakeholders is obtained through wide consultations. Historical trends also play an important role especially for non-Plan funding.

Use of formal economic methods for allocating agricultural research funds is a recent phenomenon in India. These methods are being tested under NATP for research programs at the ecoregional level. Another innovative method for resource allocation is followed in the AICRPs where ICAR and SAD agree to fund in the ratio of 75% and 25%, respectively. The locations of AICRP centres are decided based on priority ecoregions and funds are allocated accordingly.

In general, resource allocation appears to have been congruent with the distribution of production across regions. Jain and Byerlee computed a congruency index of 0.88 between value of production and resource allocation in 20 production environments for wheat.

The main discrepancy has been the strong tendency for research intensity to be higher in smaller production environments. Over time too, there is good evidence that resources have shifted in accordance with changing production conditions. In the case of wheat, this implies an increase in resources allocated for breeding for late planting, and a decrease in resources for rainfed areas, in accordance with increased cropping intensity and irrigation, respectively.

Competitive Funding

Competitive funding is gaining popularity in India. It is regarded as a powerful mechanism to direct funds to high priority areas, improve quality and accountability, and promote wider participation of research providers and innovative partnerships. There are at least five different competitive funds operating at the national and state levels to support agriculture research. Unlike other developing countries where these funds have mostly been established with donor support, several of the Indian funds were initiated with domestic resources and may therefore be more sustainable. Although these funds are increasing, they still account for a small proportion of the total public research funding (about 3%).

The Ad hoc Research Scheme of ICAR, financed by the agricultural cess on selected commercial crops, is the oldest competitive fund, supporting research in emerging areas and research to fill critical technology gaps. The Competitive Grant Programme (CGP) of NATP and Competitive Agricultural Research Programme (CARP) of Uttar Pradesh Council of Agricultural Research (UPCAR) are of recent origin and donor supported.

The competitive funds of DST and DBT support upstream research in all fields of science, including agriculture. All these funds have similar operational modalities—short-term research projects selected through peer review, and provision of funds for operating costs but not for salaries and infrastructure.

Although these funds are operating quite successfully and are in high demand, a number of issues need to be addressed. The proposals are not invited against well-defined research priorities and therefore the number of proposals is large and the success rate is low (CGP addresses this problem to some extent).

Most operate at the national level, and there is no systematic mechanism to ensure that regional priorities are addressed. This problem, coupled with weak capacity to develop competitive proposals in institutions located in less developed regions leads to a low success rate in those regions.

More effort is needed to train scientists in weaker institutions in developing research proposals. The experience of CGP has also shown that a prompt evaluation of proposals is important in attracting quality proposals. Finally, research projects under competitive grants are time-bound and therefore it is critical to have timely release of funds and efficient administrative procedures.

Sources of Funding and Fund Flows

In spite of the fact that agriculture is a state subject, ICAR funded by an annual block grant of Rs 13 billion (US$ 300.9 million) in 2000 from the central government, has the major responsibility for agricultural research. ICAR also manages funds received as grants and loans from multilateral donors and collaborative research programs funded by bilateral donors and international organisations.

The World Bank is the dominant source of such funds. Currently, ICAR manages a loan under NATP of US$180 million from the World Bank for strengthening research and extension for the period 1998-2003. A small loan was also provided for human resources development in SAUs in four states, which may be scaled up and extended to other states in a later phase.

In addition, ICAR manages the AP Cess Fund levied at 0.5% (ad valorem) on specified export commodities and accounting for about 2% of the total ICAR budget in 2000.

Finally, with implementation of a new policy on self generation of income, ICAR earns some resources through consultancies, contract research and services, sale of seed and other planting material, and royalties on research products through partnerships with the private sector. However, progress has been modest—ICAR could generate only about 3% of its total budget in 2000 through these means.

Overall, the central government provides 52% of public funding for agricultural R&E in India, which almost entirely passes through ICAR. A significant proportion of the ICAR funds (30%) are made available for extramural funding and a large proportion of these funds (87%) is directed to the SAUs.

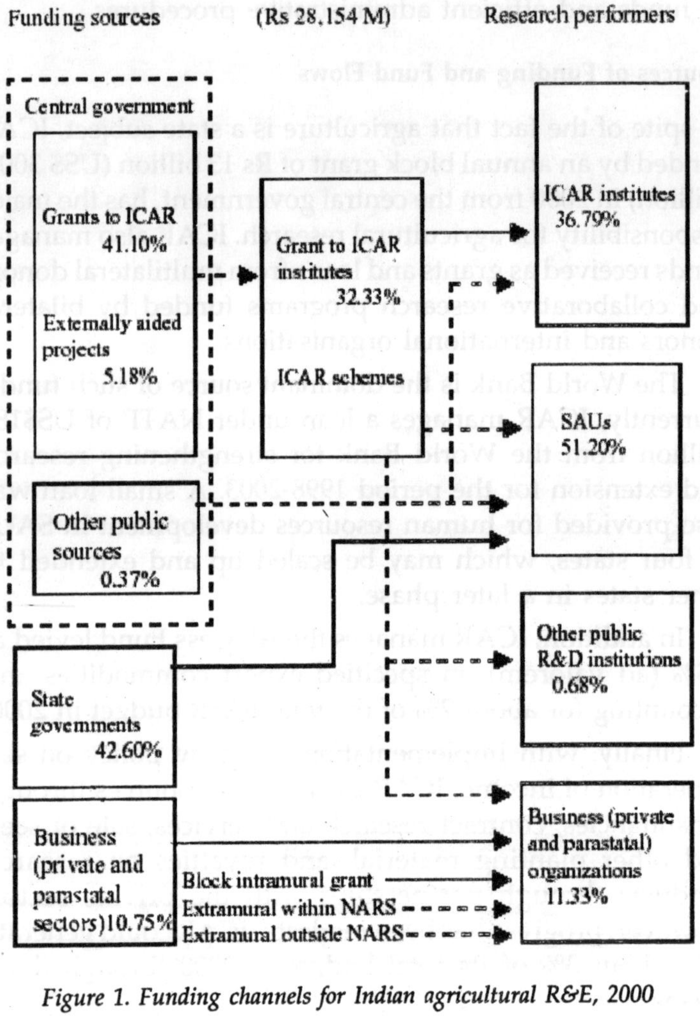

Figure 1. Funding channels for Indian agricultural R&E, 2000

Non-agricultural public research institutions and private research organisations (for profit and non-profit) obtain 7% and 6%, respectively, of ICAR's extramural funding through competitive research program and support to KVKs.In terms of funding mechanism, it is estimated that about 30% of the extramural funding from ICAR is made through the AICRPs in the form of block grants, 12% through competitive funding, 34% through donor-funded projects, 17% through grant to KVKs, and 7% as development grants to SAUs. Annual block grants from the state governments to the SAUs totalling Rs 12 billion in 2000 are the second major source of funding. Practically all of these funds are used intramurally by the SAUs.

Use of state funds by ICAR institutes does not exist. Only a small competitive fund in Uttar Pradesh is open to all research organisations located in Uttar Pradesh, including ICAR institutes. The remaining significant source of research funds is private firms which is nearly all used for intramural research and accounts for about 11% of the total. Private funding of research in public organisations is negligible. The most cited example was a research contract between ICAR and Mahyco Research Foundation for hybrid rice development in 1995.

Such linkages should increase in future because of concerted efforts being made by ICAR, but it is unlikely that these partnerships will make a significant contribution to total agricultural research efforts in the country for many years. In terms of spending (the right side of Figure 1), all ICAR institutes together accounted for 37% of the national expenditure on agricultural R&E, SAUs for 51% and the remaining 12% was spent by other public and private organisations.

Overall Public Funding for Research

India has consistently committed substantial government funds for research in all fields of science including agriculture. Total funding for agricultural R&E increased in real terms (1999 prices) from Rs 2.46 billion ($284 million 1999 P PP or international dollars) in 1961 to Rs 7.57 billion

($875 million) in 1981. This rose to Rs 25.0 billion ($2,893 million) in 2000—a ten-fold increase over the past four decades. An increasing trend is observed for both central and state funding. Funding from the states grew rapidly during the 1960s because a large number of SAUs were established during this period. Central funding has outpaced state funding thereafter and their shares became almost equal in the 1970s and the 1970s

Using simplistic assumptions, nearly three-fourths of these total R&E expenditure is spent on 'research', and in absolute terms 'research' expenditure reached Rs 16.2 billion at 1999 prices ($1,898 million 1999 international dollars) in 2000. Overall public research funding grew at 3.16% in the 1970s, 7.03% in the 1980s, and slowed to 4.61 in the 1970s These trends show strong political commitment to research in spite of a pluralistic political system, changes in governments, and shift in public investment priorities.

Table 2. Intensity of public agricultural R&E funding: All India

Indicator	*1961-63*	*1971-73*	*1981-83*	*1991-93*	*1997-99*
R&E expenditure					
Constant local currency units (M 1999 rupees)	2,697	6,576	7,892	14,335	17,885
Total expenditure (M 1999 PPP dollars)	312	760	912	1,657	2,068
Per capita expenditure (1999 PPP dollars)	0.71	1.39	1.34	1.97	2.14
Expenditure per agricultural worker (1999 PPP dollars)	2.38	6.04	6.17	8.94	9.76
Expenditure per hectare of net cropped area (1999 PPP dollars)	2.29	5.47	6.50	11.75	14.52
Expenditure as % of AgGDP	0.20	0.35	0.36	0.44	0.42
Research expenditure					
Constant local					

currency units (M 1999 rupees)	1,511	4,054	5,057	9,069	11,404
Research expenditure (M 1999 PPP dollars)	175	469	589	1,049	1,318
Research expenditure as % of AgGDP	0.11	0.22	0.23	0.28	0.31

Intensity of Research Funding

Another way to assess funding is to compute various intensity ratios such as expenditure per agricultural worker, per unit of agricultural land, and percentage of agricultural GDP (AgGDP) (Table 2). All the intensity ratios registered impressive growth over time in spite of significant growth in population, land area, and AgGDP. Agricultural research expenditure as a percent of AgGDP increased significantly during the 1960s and 1980s, but remained around 0.3 % during the 1970s This slowdown is worrying given that the average for all developing countries is 0.6 percent and 1.0 percent globally.

Part of the difference can be attributed to the relative importance of agriculture and economies of scale and scope in agricultural research, but there appears to be a clear case of under-investment as China, a country of comparable size and stage of development, spends 0.43% of AgGDP on research. Even comparing agricultural research with general science and technology research in India, ICAR received only about 10% of the total research funds of the central government in 1997.

Funding by States

The growth in real funding was highly uneven among states during the 1970s. These differences narrowed in the 1970s with steady growth in all states. The growth of total state funding increased from 1.3% per annum in the 1970s to 8.2% in the 1980s, but slowed to 3.8% in the 1970s The intensity of state funding has increased in all states since

1980s, except in West Bengal. However, there remains wide variation in the intensity between states with comparatively high ratios (over 0.4% of AgGDP) in Himachal Pradesh, Tamil Nadu, Haryana, Maharashtra, Gujarat and Kerala, and very low ratios (less than 0.2%) in the states of Madhya Pradesh, Rajasthan, Uttar Pradesh and West Bengal.

A host of factors may explain variations in the intensity of agricultural research.. Pal and Singh applied a political economy model to analyse the determinants of the level of state funding to agricultural research in India using cross-sectional and time-series data for the period 1982 to 1994. Although the results were mixed and unmeasured state-specific attributes were important, per capita state funding was found to be strongly related to per capita AgGDP indicating that states with higher income level spend comparatively more on agricultural research.

Rural literacy and the share of agriculture in government expenditure also had a positive and significant effect on research intensity. Other factors such as sources of growth in agriculture, crop diversification, and terms of trade were non-significant. That is availability of public resources and the importance assigned to agriculture play key roles in decisions on research resource allocation.

Donor Funding

USAID has been a significant funder of agricultural research. Beginning from the early 1960s, USAID supported building of the SAUs up to 1977. This was followed by other projects including a major agricultural research project in the 1970s when USAID support was at its peak. In total USAID invested some US$108 million or Rs 4.5 billion in agricultural research in India to about 1990, when support was terminated. Beginning in 1980, the World Bank became a significant supporter of agricultural research, first at the state and zonal levels, and from 1997 at the national level.

The World Bank has also supported human resources development in the SAUs from 1995 and a number of state projects have financed agricultural research, especially in Rajasthan and Uttar Pradesh. In total, the World Bank has provided US$538 million or Rs 22.6 billion to agricultural research since 1975.

One important implication of these results is that in low-income countries, such as India, donor support to agricultural research can help increase intensity levels. However, long-run funding sustainability requires India give high priority to agricultural research investment over nondevelopmental expenditures many of which are subsidies. This is particularly so when the rates of returns to agricultural research are found to be very attractive.

Private Research Funding

The recent rapid growth in private research spending in India has outpaced the capacity to track its intensity, orientation, and impact. Based on broad estimates for each subsector (seed, pesticide, machinery, livestock, and food processing) total private or business funding for agricultural research (including funding by state-owned enterprises) in India doubled from Rs 800 million (US$ 24 million) in 1985 to Rs 1,695 million (US$ 51 million) in 1995. In terms of 1999 international dollars, the funding increased from $119 million to $253 million in this period. Private research funding has grown at 7.5% compared to 5.1% in the public sector over the same period, and accounted for 11% of total funding of agricultural research in 2000 (Figure 1).

The most rapid increases in private growth have occurred in food processing, seeds, veterinary products, and sugar. More recently, there has also been strong investment in biotechnology, animal health and the poultry sector. This has been accompanied by significant growth in research expenditure by multinational companies.

Agricultural Research Funding in India

Although India has one of the largest research systems in the world, the public sector still under-invests relative to other developing countries (Table 3). In the late 1990s, India invested 0.31% of agricultural GDP in research, close to China at 0.43%, but significantly lower than the average for all developing countries of 0.62%. Industrialised countries spent a much higher figure—2.64% of agricultural GDP—on agricultural research, reflecting their relatively higher tax base, smaller agricultural sector in relation to the economy, and often politically powerful farm lobby groups.

Table 3. Public research expenditures in India relative to other regions

	Intensity (% AgGDP)	Growth rate (%/year)	
	1995	1976-85	1986-95
India	0.29	3.92	5.00
China	0.43	8.0	3.1
Latin America	0.98	4.9	2.1
Africa	0.85	2.0	0.7
USA	2.45	2.3	2.2
UK	2.63	3.8	-2.2
France	2.0	3.2	2.0
Australia	3.90	NA	1.0
All developing countries	0.62	5.4	4.1
All industrialised countries	2.64	2.3	1.6

However, India appears to be catching up with the rest of the world in terms of growth of overall spending on agricultural research. Over the period, 1986-95, the growth rate of spending at 5% was higher than for all developing countries, and comparable with the rest of Asia which had a relatively high growth rate.

One notable feature of spending for agricultural research in India is that growth has accelerated through the

1990s, in contrast with a worldwide slowdown and even decline in some countries. If India can continue this trend, research intensity should reach the average for all developing countries in the next few years.

Agricultural research does seem to be more centralised in India relative to other large countries. For example, although the federal government in the USA and the central government in India both fund around half of all agricultural research in the public sector (53 and 51% respectively), a relatively larger share of research is performed by the states in the USA compared to India (67 and 58% respectively). This is because, 39% of federal funding in the USA is transferred to the states through various mechanisms, relative to only 27% in India. However, relative to other large developing countries, India may not be unduly centralised.

The private sector share of agricultural R&D spending in India is similar to the average for Asia (Table 4) but significantly above the average for the developing world of 5%. Within Asia, private spending has increased more rapidly in India than the average for Asia, but its share is still lower than for Malaysia and the Philippines.

The composition of private spending is quite similar to that in the USA in about 1990, with spending by chemical companies and food processors as the largest share. However, the share of spending on genetics, including biotechnology, has increased sharply in the USA at the expense of spending on chemicals, and a similar trend is underway in India.

It is also important to note that the share of research in public research organisations financed by the private sector in India is low, relative to the USA where the private sector accounts for 13% of funding in USDA, and the state agricultural universities.

Table 4. Growth of private research in Asia (Millions of 1995 U.S. $s)

	1985	*1995*	*Annual growth rates*	*Private as % of total 1995*
India	26	56	7.7	13.9
China	0	16		3.2
Indonesia	3	6	6.9	6.9
Malaysia	14	17	1.9	21.0
Philippines	6	11	6.1	22.4
Thailand	11	17	4.4	11.8
Pakistan	2	6	11.0	Na
Total	62	128	7.2	10.1

HUMAN RESOURCES AND PATTERNS OF EXPENDITURES

Human Resources for R&E

Although precise and consistent estimates of scientific staff in the ICAR/ SAD system over time are not available, the number of scientists working in the ICAR/SAD system during the late 1970s was estimated at 4,189 scientists in ICAR and 14,851 scientists in the SAUs, giving a scientific strength of 19,040.

The number of scientists remained steady in ICAR during the 1990s, but decreased significantly in the SAUs and has likely further depleted in the 1990s, because of non-replacement of retiring faculty and restrictions on recruitment.

Adjusting the number of scientists by share of research expenditure relative to extension and education (for ICAR) and percent time spent on research (for SAUs), the number of full-time scientists in the late 1970s was 2,999 in ICAR and 8,132 in SAUs, giving a total of 11,131 full-time researchers in the country—a figure similar to that in the USA (Table 5). This is a substantial increase from an estimated 5,666 full-time researchers in ICAR/SAD system in 1975, and 8,389 in 1985.

Table 5. Scientific manpower and allocation of R&E expenditure in ICAR/SAD system

	ICAR (1996-98)	*SAUs (1992)*
Total number		
Researchers	4,092	17,678
Full-time researchers	2,999	8,132
Qualification (% distribution)		
Researchers with Ph.D. degree	68.8	62.6
Researchers with M.Sc. degree	31.2	35.7
Women researchers (%)	7.5	2.1
Ratio of scientific to technical and administrative staff	1:1	1:2.5
Allocation of expenditurec (%)		
Research	73.3	45.0
Education	5.2	33.0
Extension	6.1	5.0
Others (administration, publication, recruitment, etc.)	15.4	17.0

The educational qualification of Indian researchers is also impressive—more than two-thirds of researchers hold a Ph. D. degree and the rest are MS holders. The proportion of women researchers is, however, very low— 7.5% in ICAR and 2.1% in the SAUs. Scientific staff are supported by a large number of technical and administrative staff. The ratio of scientists to administrative staff is especially high in the universities at 1:2.5. ICAR and to some extent the SAUs are now attempting to balance these numbers, by downsizing administrative staff.

Resource Expenditure Patterns

In terms of research expenditures, 37% was spent by ICAR institutes, 51% by SAUs and the rest 12% by private and non-NARS public organisations. This compares with about half of the funding provided through ICAR since there is a net flow of funds from ICAR to SAUs, largely through the AICRPs. A more desegregated analysis of expenditures patterns by providers of R&E is difficult, as India has no ready means to track the allocation of overall expenditures below the institute level.

Strategic versus applied research

One way to look at R&E funding is its allocation by types of R&E—strategic, applied, and adaptive research, extension, and education—by reviewing the mandates of research providers. Basic and strategic research mainly conducted in ICAR institutes accounted for 21% of total agricultural R&E expenditure, and applied and adaptive research accounted for 53% of the expenditure. Of the remaining, 20% was used for education and human resources development, and 6% was allocated to frontline extension-related research—assessment, transfer and refinement of new technologies—in ICAR institutes and SAUs. These shares appear to be reasonably distributed although there is concern about weakening of basic and strategic research in the system.

Weakening of research capacity in the SAUs because of non-replacement of retiring faculty due to inadequate funding from the state governments, is also a matter of concern. It may be noted that the US spends about equal proportion of the resources on the basic and applied research and very little on the developmental or adaptive research. A higher allocation for the basic research in the US is justified in view of the higher private research investment mostly directed to the development research.

Favoured versus less-favoured regions

The irrigated ecoregion received high priority during the Green Revolution period primarily because of its high growth potential. This paid rich dividends in terms of a quantum jump in crop yields, but in the process, rainfed and marginal regions were neglected. This was corrected in the Seventh Plan, which gave high priority to research for rainfed agriculture. To see if past imbalances have been corrected, we compared actual research expenditure in different ecoregions with the normative allocation using the congruence rule (value of production), modified by criteria for sustainability and equity.

In fact, contrary to general belief, less-favored environments received slightly more resources than those justified by the efficiency criterion, even after inclusion of natural resources and equity concerns which favored allocation to rainfed areas. These very broad observations are supported by analysis of resource allocation for a specific commodity, wheat, by Byerlee and Morris who used the number of field experiments as a proxy for investment by agroclimatic zone. They found that in spite of the predominance of irrigated wheat and its high research pay-offs, there was no evidence of underinvestment in marginal environments.

This was further reinforced by a detailed study by Traxler and Byerlee which showed that rainfed and hill environments accounted for 30% of resource allocation to wheat breeding research although these environments only produced 12% of total Indian wheat production. More revealing is the estimate that these research programs for rainfed and marginal areas only produced 1.3% of the benefits generated from wheat research in India during the post-Green Revolution period, 1976-93.

Sub-sectors and commodities allocation

Data on research expenditure by sub-sector and commodity are only available for ICAR, including research expenditure on AICRPs in SAUs. Together these represent a large share (67%) of total research expenditures in the ICAR/SAD system. In ICAR, crop research received the highest proportion of research expenditure followed by animal sciences and natural resource management.

Normative allocation pattern based on value of production indicates that crop research should get 51% of resources, followed by animal science 28% and horticultural crops 21%. Nevertheless, both livestock and horticulture are high growth sub-sectors that might justify slightly more resources than indicated by value of production, although in the case of livestock research, this might be counterbalanced by the fact that livestock research is known to be less location specific with higher spillovers.

ACCOUNTABILITY AND RESEARCH IMPACT

A number of monitoring and evaluation mechanisms have been put in place at the national, system, institute and project level to ensure relevance of research and accountability in the use of public funds. At the national level, the Planning Commission and government committees monitor the progress and achievements during preparation of annual and five-year plans.

At the regional level, there are eight regional committees comprising representatives of ICAR, SAUs and government departments to assess the status of agricultural research in the region (covering several states), and make recommendations on research priorities. At the institute level, there are management and research advisory committees to oversee administrative and financial matters, advise on research programmes, and monitor their progress. In each ICAR institute a staff research council with representation of external experts evaluates research projects.

A more substantive external review of each ICAR institute and each SAD is done by an external review team every five years. The review process covers organisational, management, scientific and other matters relating to effectiveness, efficiency and relevance of the institute. In addition, for SAUs, there is a committee to determine the norms for accreditation and financial assistance from ICAR and to periodically assesses performance of SAUs against these norms.

Through these mechanisms, accountability for the use of public funds is high. However, questions are often raised about the effectiveness and impact of the research system. This is in spite of the fact that the Indian research system successfully led a series of technological revolutions in the agricultural sector.

Many studies have empirically examined the impact of agricultural research in India by estimating internal rates of return to investments. Most have analysed returns to crop research, individually or for the subsector as a whole. Although there is considerable variation, the average return was about 70%, with a median value in excess of 50%. Interestingly, there is no evidence that the rate of return has declined since the Green Revolution.

The studies have also shown that returns to public research investments have been higher than those for public extension or private research. These results provide a convincing case for enhancing public funding to agricultural research. This point has been made repeatedly by research leaders to build the case for the higher budget allocations, particularly during the preparation of the five-year plans. These efforts have achieved some success as demonstrated in the steady rise in public funding to agricultural research over the past two decades, in spite of fiscal discipline adopted by the government during the 1970s

Table 6. Internal rates of return (%) to research investment in India

	Aggregate analysis	*Analysis for individual crops*	*All*

Mean	75.4	69.9	71.8
Median	58.5	53.0	57.5
Minimum	46.0	6.0	6.0
Maximum	218.2	174.0	218.2
Number of studies	10	18	28

Finally, it should be noted that high aggregate rates of return may be hiding considerable inefficiencies in the Indian public research system. Traxler and Byerlee analysing rates of return to 20 wheat breeding programs across 50 research stations, found that although the aggregate rate of return to wheat improvement research in India over the period, 1978-1991 was estimated to be 55%, eight programs had a negative rate of return when spillins were taken into account. Research output was concentrated in the two strongest programs which generated 75% of all benefits, even though they claimed just 22% of research resources. Clearly there is considerable scope for increasing the overall return on research investment by redirecting money from unproductive research programs.

EMERGING POLICY ISSUES

Agricultural research policy must respond to a changing agricultural, scientific and economic environment. In the industrialised countries, agricultural research reforms originated from the declining importance of agriculture in the economy and the rapid increases in private research investments. These reforms included separating research funding from research execution, encouraging contestability of funds through competitive mechanisms, improving accountability of research institutions, and shifting near-market research to the private sector.

The new paradigm underscores pluralistic institutional structure, new sources and mechanisms for research funding, organisation and management reforms of public institutions, and management of intellectual property.

These same reforms are generally proceeding more slowly in developing countries, where there are a large proportion of small scale farmers and the public sector still dominates the research system. Thus the focus of research policy should remain on improving efficiency of the public research system and encouraging participation of the private sector where possible.

Balancing Multiple Research Objectives

The Indian NARS must find a balance among multiple objectives, ranging from traditional food security objectives, to emerging demands to serve a more market oriented economy, meet the needs of more sophisticated consumers, and preserve the environment. Striking a balance between these objectives has major implications for organisation of research, prioritisation of the research agenda, and management of intellectual property.

Since there are increasing demands on the public sector to provide technologies with characteristics of 'public good' and that address market failures in addressing social and environmental concerns, public research investment in India needs to close the gap with the global average of one percent of agricultural GDP. Also, public research institutions must work closely with key stakeholders to define priorities that address multiple objectives, employing formal research prioritisation approaches. This is extremely important when the system is large in size, objectives are conflicting and clients are poor in articulating their research needs. A starting point in this direction would be to carefully track current resource allocations and make necessary adjustments to reflect changing priorities.

The distinction between the roles of the center and the states in agricultural research has become blurred over time. In practice, SAUs should have primary responsibility for applied and adaptive research to meet local demands in their respective states, and ICAR should take the lead in strategic research that is relevant to several states, and in those applied research areas where states will tend to under-invest due to spillovers. However, SAUs are generally starved for operating funds and now largely depend on ICAR. A shortage of funding in the SAUs has had adverse effects on human resources development, research infrastructure, and linkages with farmers. There is an urgent need to sensitise policy makers at the state level to the pay-offs to investing in research.

At the same time, the central government might develop a funding formula that supports the weaker states, but provides incentives to stronger states to increase their funding. A key role of central research is to generate spillovers to enhance efficiency in state research programs. In some areas, especially crop breeding spillovers are pervasive. The AICRPs provide a mechanism for facilitating such spillovers.

More Pluralistic System

The modern concept of a NARS emphasizes a pluralistic system of research providers that recognises the comparative advantages of different providers, and complimentarity that can be achieved by forging close linkages between different actors. The leadership of ICAR has noted these requirements and has taken a number of initiatives to promote such linkages. However, but effective implementation needs greater awareness down the line. In particular, the growing role of private research and the implications for public institutions are not widely appreciated. Where the private sector can efficiently provide near-market research services with scope for appropriation of benefits, the public sector should be prepared to withdraw and play a complementary role.

Private research is stimulated by strategic research support from the public sector, and there are many areas where public-private linkages can enhance the effectiveness of both sectors. Enabling institutional mechanisms, especially IPR protection and capacity within the public sector to manage partnerships, can help develop and sustain these linkages.

Sustainability of Research Funding

The Indian public NARS has been relatively successful in increasing government funding for R&E. However, the current funding situation is not sustainable for a number of reasons. First, increased funding has not matched the continuing expansion of the number of R&E institutions, resulting in a steady increase in the share of salary and overhead expenditures at the expense of operating expenditures.

In ICAR, the salary to operational expenses ratio has increased to 70:30 against a target of 60:40 and the situation is even more serious in the SAUs. Second, although competitive funding has increased, it still accounts for a low share of total funding. Because competitive funding has the potential to enhance accountability, quality and efficiency of the system despite somewhat higher costs in terms of overheads and time of scientists, a higher share of funds should be gradually shifted to competitive funding. Of course, regular block grants must continue in order to maintain and upgrade research infrastructure, and strengthen basic and strategic research.

Finally, new resource generation opportunities such as payments for services by farmers growing high value crops, income generation through commercialisation of technology and services, and contract research with the private sector are emerging and should be tapped. ICAR has set a target of 25% of budget share from these sources by 2020. This will require development of capacities in IPRs and business skills in public research organisations. ICAR has already developed such a policy, and the government has offered matching grants for self-generated income as an incentive.

Challenges of Modern Science

Although India has developed relatively good capacity in new areas of science, especially biotechnology, these have raised a number of challenges—development of research capacity, biosafety and IPR regulations, and management of public dialogue on controversial issues. Establishment of biotechnology capacity is relatively capital and human resource intensive. Although it is expected that the private sector will be an active player in biotechnology in India, the public sector will have to play a dominant role, especially for non-commercial agriculture. Therefore, mechanisms to access proprietary technologies by using resources in the public sector, such as germplasm, as bargaining chips and segmentation of markets deserve special attention. Also, given the number of public and private institutions involved, there is much potential for forging publicprivate linkages to enhance overall impacts. These include sharing of cost and benefits, joint ventures, and management and ownership of intellectual property.

Advances in biotechnology have also blurred the differences between general sciences and the agricultural sciences, requiring close linkages with general science and technology providers. This is more so when major responsibility for promotion of biotechnology in India rests with DBT in the Ministry of Science and Technology.

Given the current debate on biotechnology in India and elsewhere, effective biosafety regulations must be in place that are credible, cost-effective and properly coordinated. This is the single major constraint to application of transgenic technology in India, which still has only just released the first product (Bt cotton), despite many years of research and many products in the pipeline. Finally a dimension often neglected is the provision of information about these new technologies to farmers.

Organisation and Management Reforms

The public sector in general in India suffers from centralisation and bureaucratisation that imposes high transaction costs at all levels. Despite having a certain level of autonomy, the research system is no exception. Although ICAR recognises these problems and has initiated a number of organisational and management (O&M) reforms, there are still important gaps as well as problems in their implementation. First, institutional rigidities imposed by commodity and disciplinary boundaries restrict the flow of information between hierarchies and organisations in a large system such as India's. The decision to review the functioning of the AICRPs—originally established to forge interdisciplinary and inter-institutional research—was an important step toward addressing these rigidities. But much remains to be done to decentralise and devolve power before transaction costs can be reduced to acceptable levels for efficient research management.

Second, there is a growing problem in the quality of scientific human resources owing to inbreeding in the system, especially in the SAD system, and weakening of global scientific linkages. In the 1960s and 1970s, a significant proportion of scientists were educated abroad and Indian scientists were generally well integrated with regional and international networks. This situation has deteriorated significantly with scientists often working in the same institution in which they receive their PhD, and with professional isolation of many scientists. This trend must be arrested through assessment of human resource needs and use of foreign grants and loans for human resources development, and to support participation in international scientific networks and other initiatives. Advances in information and communication technologies also have much potential to foster such linkages and improve access to international literature and scientific data bases. At the same time, performance-based evaluation of scientists that is linked with incentives and the reward system is long overdue.

Third, research institutions require much improved accountability through institutionalisation of objective and transparent evaluation mechanisms for planning, monitoring and impact assessment of research. Proliferation of research programs has meant that many programs serving small states and agro-ecological zones are inefficient. Much of the inefficiency found in the Traxler and Byerlee study is due to research programs serving small ecologically- and politically-defined markets, so that even if they are productive in terms of technologies produced, they are only used in a small area. Resource allocation needs to be linked to research planning based on 'bottom up' approaches involving relevant stakeholders and feedback from monitoring and impact assessment. Implementation of such processes has been attempted several times, albeit with varying degrees of success. A prerequisite for its effectiveness is to link planning, monitoring and evaluation with funding decision and with performance evaluation at various levels—the system, institute, project and scientist.

Finally, although successive review panels of ICAR have raised these various concerns and proposed recommended changes, past attempts at reform failed due to lack of financial flexibility and autonomy of ICAR. A package of reforms aimed at enhancing autonomy, improving decentralisation and devolution of power, and improved financial management through project based budgeting is required. Both ICAR and SAUs should commit themselves to such reforms. Support of high level policy makers at both the central government and state government levels is needed to implement a this far reaching reform agenda.

Technology Transfer

It is generally agreed that play-offs to agricultural research could be much higher with a stronger research-extension interface. The weaknesses of the current system can be attributed to a number of factors:

— adaptive research and technology transfer is considered to be a less challenging task, and therefore, not many scientists are attracted to it;

— scientists working for technology assessment and transfer are at a disadvantage since the number of publications dominates performance evaluation criteria;

— most scientists lack skills to assess farmers' research needs and design appropriate technologies; and

— scientists lack operating expenses for on-farm research.

In addition, supply-driven extension approaches focused on the public sector in India are long overdue for drastic overhaul. Improved accountability to clients through incentive systems in the research system and piloting of more pluralistic and demand-driven extension systems are now being given higher priority as a way to speed technology transfer.

REFERENCES

Alex, G., *USAID and Agricultural Research: Review of USAID Support for Agricultural Research:1952-1996*, Agricultural Research and Extension Unit, the World Bank, Washington, 1997.

Alston, J.M., J.E., "Christian, and P.G. Pardey. 1999. Agricultural R&D investments and institutions in the United Sates", *In J.M. Alston,* P.G. Pardey and V.H. Smith (eds.), *Paying for Agricultural Productivity,* Johns Hopkins University Press, Baltimore and London, 1997.

Byerlee D. and M. Morris, "Have we underinvested in research for marginal environments?" The example of wheat breeding in developing countries, *Food Policy*, 18: 381-393, 1993.

Carney, C., G.J. Gill, and S. Pal, *Improving competitive agricultural research funding in India*, Policy Brief 12, National Centre for Agricultural Economics and Policy Research, New Delhi, 2000.

Desai, B.M., *Agricultural Development Paradigm for the Ninth Plan Under New Economic Environment*, Oxford & IBH Publishing Co., New Delhi, 1997.

Evenson, R.E., C. Pray, and M.W. Rosegrant, *Agricultural Research and Productivity Growth in India*, Research Report 109. Washington: IFPRI, 1999.

Fan, S., P. Hazell, and S. Thorat, *Linkages between Government Spending, Growth, and Poverty in Rural India:* Research Report No. 110. IFPRI, Washington, 1999.

6

Development of Agricultural Research and Extension

Irrigated areas producing the major cereals (rice and wheat) have long served as a breadbasket to India's burgeoning urban population. This role was greatly reinforced by the Green Revolution, when fertiliser-responsive, short-straw varieties were taken up rapidly. There have been occasional instances of rapid technology-based growth in crop production in the more reliably rainfed areas, such as in oilseeds. On the whole, conditions between irrigated and much of the rainfed land differ substantially, and it would be naive to suppose that a second Green Revolution might occur in rainfed areas. Briefly, the rainfed areas are disadvantaged in a number of ways:

— Widely diverse and, in some cases, environmentally fragile agro-ecological conditions, with irregular rainfall and nutrient-poor soils, contribute to the need for risk-averse approaches to farming.

— The farming population has limited capacity to articulate their needs, unless supported by participatory approaches.

— A wide diversity of varieties, cropping patterns and farming practices makes it difficult for research to respond specifically to these.

— Weak infrastructure makes it difficult to engage the private sector with all but high-value, easily transported products.

The above conditions have two broad implications for research:

— It is difficult to replicate farmers' conditions on-station, so that farmers' opinions and knowledge have to be drawn upon in two ways: by having them visit research stations and give their opinions at an early stage on the technologies being tested; and by conducting more adaptive testing under farmers' conditions.

— Farmers cannot be assumed to be motivated by profit maximisation alone: much stronger effort is needed than in irrigated areas to understand the influence on their farming practices of risk-aversion, and of other socio-economic and cultural factors, including the role of women.

NEED FOR SOCIAL-SCIENCE SKILLS

Farming conditions in many parts of the rainfed areas are complex, diverse and risk-prone. The need for social-science inputs into the research process, and into related organisational and institutional change, is determined fundamentally by the fact that these conditions cannot easily be replicated on-station and subjected to reductionist observation. Compared with irrigated areas, a greater part of the effort towards technical change must be made on-farm, under existing farming and natural resource-management systems, and in the context of prevailing socio-economic and behavioural modes and conventions.

Rainfed areas differ fundamentally from irrigated areas in this respect, and in the amount of risk involved in agricultural production. On-station experiments typically eliminate or minimise risk, so for rainfed areas may result in recommendations for input levels and practices which farmers find unacceptable. Social-science skills are essential to understand farmers' risk-averting practices and to design technology appropriately for them.

However, social-science skills are valuable not only because they may lead to technologies that yield productivity gains. They may also contribute to technological changes with acceptable levels of distributional equity, and to the sustainability, both institutional and environmental, of those changes. Sustainability and equity are, of course, concerns relevant also to irrigated areas, especially for instance, in the rehabilitation of micro-watersheds, that they have unique characteristics in rainfed areas. The CGIAR has recognised the contribution of social sciences to improved technology design through a combination of inputs at project, policy and institutional levels.

EXTENSION AND TECHNOLOGICAL CHANGE

The findings of research reach farmers in the form of improved technologies. These may be either hardware or software. The introduction of improved technology by farmers is called "innovation". However, this is not a simple process. Successful innovation depends on:

— the appropriateness of the technology to farmers' conditions;

— the availability of appropriate information and inputs through which the technology is conveyed; and

— favourable supporting conditions, such as credit, market, and so on.

The more the farmers innovate, the more effective research and extension have been. This necessitates improvements in procedures for varietal testing and release, improved mechanisms for seed delivery, and client-orientation in deciding on plant breeding priorities. Attempts to improve the effectiveness of public-sector research and extension have focused too strongly on components of research or extension, and have lost sight of what needs to be done to put in place the preconditions for innovation. There have been rare exceptions to this narrowness of approach in respect of rainfed farming, such as the Oilseeds Production Mission. We recognise the success of this mission-mode approach, but feel unable to recommend it as a general practice for two main reasons:

— The co-ordinated attention it demands of researchers, extension personnel, the rural banking system and policy makers is unlikely to be sustained over more than a very few crops for the necessary 5-10 year period.
— The commodity focus of the approach distracts attention from chronic constraints attributable to a fragile and diminishing resource base.

Efforts to improve the resource base are channelled especially through the rehabilitation of micro-watersheds. Such rehabilitation reduces soil and water runoff, improves percolation, increases the availability of fodder and water for irrigation, and so results in the introduction of new crops and varieties. Farmers are unlikely to be familiar with technologies that will enable them to take full advantage of these possibilities. Research and extension services can play a major role in providing these from outside. The uptake of farm-level technology could also be better enhanced through improved provision of the genetic resources used by farmers.

ORIGIN AND STRUCTURE

The apex organisation for agricultural research in India, the Indian Council of Agricultural Research, currently provides core funding to 45 Central Research Institutes, 30 National Research Centres, four National Bureaux, 10 Project Directorates, and their 165 regional stations. The ICAR also funds 86 All-India Co-ordinated Research Projects, which have some 1100 co-operating centres, and a further 16 other projects or programmes.

The ICAR also promotes research, education and extension through 28 State Agricultural Universities and one Central University by providing financial assistance in various forms. The ICAR also maintains a network of 261 Krishi Vigyan Kendras and eight Trainers-Training Centres along with eight Zonal Co-ordinating Units. The major research activities of the ICAR and SAD units are elaborated in Table 1.

Table 1: Major research activities of ICAR and SAUs

Component	*No.*	*Main activities*
ICAR		
National Research Institutes	4	Basic and strategic research of national importance, education, manpower training
Central Research Institutes	41	Commodity- or resource-specific basic and strategic research, with divisional setup, education
National Bureaux	4	Conservation and exchange of germplasm and natural resources
Project Directorates	10	To fill critical research gaps in All-India Co-ordinated Research Projects, research co-ordination
National Research Centres	30	Commodity-, resource- or discipline-based strategic research in mission mode
All India Co-ordinated Research Projects	86	Co-ordination of commodity- or resource-specific research in different zones
Agricultural Universities		
Central Agricultural University	1	Region-specific applied research and education
SAUs 28		Applied research for the state, education
Zonal Research Stations	120	Adaptive research for the zone

Through these programmes and projects, ICAR also provides approximately 18% of the research funds available to the State Agricultural Universities, and a disproportionately high level of operating expenses for research, given that the much of the SAUs' revenue from the states goes into salary payments. It is impossible to give precise statistics on the proportion of ICAR's budget that is allocated to research for rainfed farming: several of the institutes and programmes focus exclusively on rainfed farming, but so also do substantial parts of the mandate of other institutes. However, informal estimates suggest that around 30% of ICAR's resources are devoted to rainfed farming.

With financial support principally from the Rockefeller Foundation and USAID, the research services were organised to emulate the strengths of the US Land Grant model as it operated in the two decades before the Second World War. This entailed a strong local and practical focus by the state-level agricultural universities in both training and research; strong interaction between the state and federal levels, in which the states feed into and benefit from the more generic types of research conducted by organisations of central government; strong links between research and extension, including feedback to research of farmers' requirements; and a dominant ethos of client-orientation across the entire system.

Although the US model was worthy of emulation, it is clear that two key contextual features that contributed to the success of the US Land Grant system remained largely absent from India. These are: strong, articulate farmers' organisations capable of making demands on the research system, and a clear role for the private, commercial sector in marketing inputs which incorporate new technology, such as seeds and agro-chemicals. Numerous reports testify to the substantial successes of the research system for irrigated areas. The characteristics of these greatly facilitate the design and introduction of new technology, namely:

— homogeneous agro-ecological conditions, easily replicated on research stations;
— articulate, commercially oriented farmers, operating under conditions of minimal risk and prepared to visit research stations and convey their needs to researchers, give feedback on technologies already tried, and so on;
— large, homogeneous markets, allowing easy and profitable penetration by the private, commercial sector in marketing inputs and outputs; and
— good physical infrastructure, allowing low-cost communications and transactions.

Approximately 68% of India's net cultivated land is currently rainfed. It supports some 360 million people, rising to perhaps 600 million by 2020. Even after the realisation of India's full irrigation potential by 2013, it is estimated that around 50% of India's total cultivable area of 142 million ha will remain rainfed. By 2020 this may need to support some 600 million people. Clearly, science will have to generate-and extension services will have a key role in disseminating-the productivity-enhancing technologies that will be necessary to improve the livelihood of this burgeoning population. The conditions outlined above which facilitated rapid technical change, in high-potential irrigated areas, are largely absent from rainfed areas:

— Rainfed agro-ecological conditions are widely diverse, even within small areas; irregular rainfall makes conditions risk-prone in many areas; and nutrient-poor soils provide a weak basis for conventional agricultural intensification. The environmental and institutional sustainability of technical change are of special concern, as is distributional equity, especially in areas where weaker sectors have only limited access to natural resources.

- Farming population have long been outside the political and economic mainstream. Levels of self-confidence and of formal education are low, making it difficult for farmers to articulate their technology needs.
- The diversity of varieties, cropping patterns and farming practices in use means that markets for any individual input are small and fragmented. This, together with poor transport and communications infrastructure, acts as a powerful disincentive for the private, commercial sector to become involved in input supply.
- Difficult living conditions in many rainfed areas make public-sector staff reluctant to stay in post there, resulting in exceptionally high turnover of staff and a high proportion of vacant posts.

Farmers' responses to these conditions include a number of risk-avoiding practices. They avoid using high levels of purchased inputs in case crop failure leads to major financial losses. They rely on well-tried early-maturing varieties that meet both their grain and fodder needs. They crop in mixtures or relays. They rely heavily on the use of biomass from off-farm sources. They integrate crop and livestock production closely. There is no comprehensive review detailing the proportion of technologies generated by rainfed-farming research that have been taken up, or comparing this with irrigated areas. However, there is substantial, informal evidence of low uptake at some stations we visited. Hard data are available in one specific area: the replacement by the research and extension system of existing crop varieties by new ones.

A well-functioning research service is aware of the types of crop varieties preferred by farmers in different agro-ecological zones, and will use these as a basis for breeding new varieties. If researchers are able to target farmers' requirements accurately, then existing varieties will be replaced with new ones every few years, and the average age of varieties grown in any one season, weighted by area planted, is low. In highly efficient agricultural systems, the average age is only a few years. Thus, in the UK, wheat-breeding services produce adoptable new varieties with such frequency that the average age of cultivars in use is only approximately three years. Given the more difficult conditions in India, the average age of cultivars might be expected to be somewhat higher, say, 8-10 years. An ICAR report suggests that the continuing low uptake of new technologies in complex, diverse and risk-prone rainfed areas is attributable, among other things, to:

- Insufficient study of the farming situations, and inadequate perception of farmers' circumstances and their needs by scientists.
- Widely differing conditions on the research stations (where research is conducted) as compared to farmers' fields.
- Excessive discipline-orientation instead of problem-orientation of the scientists.
- Inadequate farmers' participation in formulation of research and transfer-of-technology agenda.
- Recommending technology based on the results at the research station without adequate and proper testing in the farmers' fields under the farmers' resource and risk situations.
- Lack of appreciation to distinguish science from technology.

Two notes of caution should be sounded in relation to current enthusiasm for conducting a higher proportion of research on-farm:

- We should not overlook the fact that science has an important role to play in broadening the range of options available to farmers, and that certain on-station work at the strategic or applied levels of research remains essential. For instance, some initial screening of lines available to plant breeders is more efficiently done on-station than on-farm. What is important, however, is that the criteria used for selection should reflect farmers' own conditions and preferences. In all cases, the important requirement is that the interface between what farmers require and what science offers has to be managed well.

— The term "farmer participatory research" has come to mean very different things to different types of organisation. Many NGOs, for instance, pursue intensive, long-term types of participation which aim to empower farmers to recognise the conditions underlying their poverty, draw on their own resources as far as possible to improve these conditions, and draw in external resources where necessary. NGOs' freedom to mandate themselves to work intensively in a few villages makes these empowering types of participation feasible. By contrast, public-sector services such as research and extension have to be spread much more thinly over a wide area. The concern within the public sector is therefore largely with functional types of participation which do not directly take on the major task of empowerment, but aim for the more modest goal of making public-sector services more relevant to farmers' needs.

Income is a crude indicator, yet it is clear that, even within rainfed areas, higher-income farmers tend to rely less on common-pool resources for the livelihood, and so can operate on a largely individualistic basis. Functional forms of participation work well under these conditions. By contrast, the lower self-confidence of poor households and their heavier reliance on common-pool resources place a premium on group action, and on the skills necessary to elicit what types of technical changes are consistent with their needs. There is still scope for functional types of participation in these contexts-for instance, in the introduction of new crop varieties. However, numerous technical challenges, especially those concerned with natural resource management, can only be addressed if more empowering forms of participation based on group action are undertaken.

In such cases, there is substantial scope for research and extension services to link with the types of organisation, such as NGOs, which claim a comparative advantage in these areas.

Three major implications for research arise from these conditions:

— Research conducted on-station is unlikely to reflect the agro-ecological and socio-economic factors faced by farmers. It therefore tends to generate recommendations for varieties which farmers find inappropriate, and for levels of purchased inputs which farmers find too high and too risky. A high proportion of experimentation for these areas therefore should be conducted on-farm and with farmer involvement.

— Researchers need the skills to engage with farmers and elicit their needs, and they need to engage with other organisations capable of interpreting and expressing farmers' needs.

— There are powerful arguments-relating both to research and extension-for organising farmers into groups to deal with:

  — the introduction of certain technologies (eg, integrated pest management);

  — the introduction of new practices for the management of common-pool resources (eg, soil and water, grazing land);

  — interactions with the market (especially the

purchase of inputs such as seed which contain new technology); and

— the articulation of technology needs and the acquisition of technical advice from research and extension services or other sources.

The Johl Committee of ICAR also highlighted the need and scope for substantial farmer participation through farmers' associations in developing appropriate research programmes. These were necessary to take into account the complex knowledge, ecology and social environment of farmers.

Organisation of Research Services at the State Level

India has 28 SAUs operating from 65 campuses, and a further four ICAR institutes with "deemed university" status. Although initially conceived as single-campus institutions, the number of campuses per SAD now ranges between two and seven. Agro-ecologically specific "outreach" research is conducted by 120 Zonal Research Stations falling under the authority of the SAUs, and these in turn are supported by over 600 research centres and substations. The National Agricultural Research Project (NARP) initiated in 1978 strengthened research infrastructure in the Zonal Research Stations substantially.

The SAUs have state-wide responsibility for teaching, research and extension education. These were intended to integrate closely across all three functions and to embody an ethos of service to the rural community, through such actions as specialised training for rural youth outside conventional degree programmes. SAUs were expected to conduct primarily applied and adaptive research, leaving basic and strategic research to the ICAR system.

In practice, the principal research link between the ICAR system and the SAUs has been via the All-India Co-ordinated Research Programmes. The SAUs have taken up responsibilities in relation to teaching, research and extension education to varying degrees, and the facilities established for training extension staff in particular have been poor. The weaknesses of the SAUs:

— continuing lack of location- and problem-specificity in research outputs;
— excessive fragmentation of research infrastructure, preventing critical mass in any one place;
— ineffective co-ordination among universities within a state, and among campuses within a university;
— establishment of degree programmes with extremely poor infrastructure and a large intake of students;
— excessive in-breeding: in several universities up to 90% of the academic staff obtained all their degrees in the university to which they were subsequently appointed;
— weak research-extension linkages;
— unclear definition of roles between SAUs and the line departments in extension education.

A study by Manage on farmer participation in research and extension revealed that the research programme-planning process of many ZRS was not relevant to the farmers' needs, nor have the researchers been able to embrace problem-solving adaptive research mode. A review of the effectiveness of decentralisation in research funding and prioritisation draws on evidence from Bihar and Tamil Nadu. It argues that if decentralised systems are working well, "researchers should be receiving feedback from farmers and extension agents about priority problems for new research and about the effectiveness of past research products, and there should be evidence of significant on-station and on-farm trials on local problems".

In reality they find that research priorities are set centrally and in a top-down and mono-disciplinary fashion, and they reflect few of farmers' priority concerns. Further, there is little evidence of feedback to research via extension services; those with the most interaction with research tend to be senior extension staff, who in fact have least contact with farmers. The study also revealed the perception of researchers regarding the expected benefits from on-farm and on-station research, calling greater attention for improvements in incentives and reward structure.

Case studies made while preparing the National Agricultural Technology Project on horticulture research in Maharashtra reached some persuasive conclusions. These included: a low standard of research priority-setting; the duplication of effort among universities, and between universities and other institutions and programmes within the state; excessive SAD reliance on ICAR funding, resulting in distortions in research agenda; and lack of operational funds for research.

A nation-wide study on citrus research by a parallel team discovered that no fewer than 36 research units of different types were conducting research on citrus. This involved 45 scientists from ICAR and 41 in the SAUs-but only 22 ICAR and 11 SAD scientists did such research on a full-time basis. Of the 122 research activities being conducted, 93 were classified as "applied". There was hardly any "strategic" or "basic" research in the ICAR system, and hardly any "adaptive" in the SAUs. Apart from poor communications among researchers, the case study also indicated the absence of research-programme planning within the framework of national or state policy objectives, the rarity of cost-benefit analysis to assess potential impact and adoption, and the lack of reviews to assess the continuing relevance of on-going programmes.

The co-ordinated projects have become an important instrument of collaborative agricultural research in India's different agricultural institutions. There has been criticism of the AICRPs: some work currently under AICRPs could be conducted more effectively and cheaply at single locations than in a network context, there are substantial gaps in some of the AlCRPs through significant on-station and on-farm trials on non-compliance of selected centres, and problems of non-comparability of data are common. An ICAR committee on AlCRPs reviewed the relevance and continuance of all existing projects, and made recommendations for improving their structure and management.

FINANCE FOR AGRICULTURAL RESEARCH

Current levels of finance for agricultural research and education have reached approximately 0.49% of agricultural GDP. But the share of agricultural research has lost ground from 20% of all research funded by central government in 1960-80, to under 12% by the late 1980s; major increases in allocations have gone to the Department of Science and Technology and to space research. In financial year 1993-94, some Rs 424 crores was spent on the ICAR headquarters and central institutes, and around Rs 526 crores on SAUs.

Some 18% of SAD research expenditure came from the ICAR, but ICAR's contribution to agricultural education at the SAUs was halved between the Sixth and Eighth Plans. ICAR's expenditure on the KVKs increased fourfold over the same period, with the result that in the Eighth Plan, slightly more was allocated to the KVKs (12.3%) than that to agricultural education at the SAUs (11.9%). It has been argued that the shares of Plan expenditure on agricultural research and education among states bear no relation to their agricultural population, the number of farm holdings or even the total production of foodgrains in the various states.

Certainly, the northern region of India, which is dominated by irrigated production, consumes almost half of the total allocation, with approximately a quarter going to the southern region and the remaining quarter divided between eastern and western regions. There are major concerns over the decreasing operational budgets for research at both ICAR institutes and SAUs:

"During the last 14 years, the share of establishment costs in the total expenditure increased from 26.33 to 59.22%. Non-recurring expenditure fluctuated within a narrow range of 16 to 19%. However, operational and maintenance expenditure decreased from 57 to 22%".

The same source argues that the situation is considerably worse in many SAUs than in ICAR institutes. Special funds, the largest of which is the Agricultural Produce Cess Fund, are seen by some of the institute directors interviewed as a means of increasing the amount of operational funds for research. Indeed, this source provided an additional Rs 8.6 crores of funds in 1993-94. Fund utilisation from the AP Cess Fund Scheme by scientists of ICAR and SAUs has been poor. This is due to long and uncertain delays in processing applications, tardy fund disbursements, inadequate control by the principal investigator on the funds he has brought through the scheme, and low motivation on the part of scientists to earn additional funds. The ICAR has recently issued fresh guidelines on the AP Cess Fund Scheme, addressing many of the above issues. Important among these is the decentralisation of financial powers to the principal investigator. However, the actual implementation of these new, liberal guidelines is uneven.

Several people in their interview mentioned that scientists bringing project funds into the universities face difficulty in drawing on those funds when they need to; in some cases, university finance officers transfer the funds into what are perceived to be other pressing requirements. To counteract this, it was recommended in the UP Diversified Agricultural Support Project that funds from the proposed agricultural research fund should be held in a separate bank account in the name of the project leader.

Several institute directors during interview noted that the problem lay not only in limited budgets, but also in the timing of their disbursement from the ICAR headquarters. For instance, in CSWCR&TI, between 55% and 65% of Plan and non-Plan funds for operational expenses were received in the final quarter of the financial year, making it difficult to maintain the necessary profile of research throughout the year.

To augment the financial resources of the institutes in a substantial manner, ICAR came up with a revolving fund of Rs 15 crores out of the AP Cess Fund, in which a one-time grant is given to ICAR institutions, SAUs and KVKs. The amount is recoverable from the fifth year onwards in equal instalments. This provision is presently of limited value for long-cycle activities such as horticulture or forestry. The financial rules governing the purchase of inputs and sale of produce also cause problems in implementing this scheme. The intention of revolving funds is mainly to generate resources or profit that should make repayment possible. The directors argued that these activities should be freed from the normal rules of purchase and sale.

ORGANISATION AND MANAGEMENT OF RESEARCH

A number of changes have recently been, or are being, introduced to the research system. These are of generic nature and not specifically geared to the needs of rainfed farming. However, these can be expected to have a positive influence on the performance of research for rainfed farming. These changes include:

— The provision for self-generation and retention of earnings by individual ICAR institutes. The level of these as a percentage of overall budgets varies from one institute to another. The implicit intentions underlying this provision are that institutes should become more strongly client-oriented; that staff reward-systems should be shifted to reflect their performance in meeting clients' needs, and that the availability of these funds should increase institute directors' flexibility to meet perceived needs.

— The provisions of the Johl Committee report on partnership, resource generation, training, consultancy, contract research/contract service and incentive and reward systems. Basically, this report recommends closer partnerships between the ICAR and the private sector, and the introduction of certain practices characteristic of the private sector in applied and adaptive research, and in the provision of related services and training. It is envisaged that ICAR's products and processes will be much more widely patented than in the past, that royalties will be obtained, and that charges will be made for advice, testing services and training. Whilst the ICAR will remain mandated to safeguard the interests of farmers, partnerships and practices of this kind will allow it to generate the resources necessary for modernisation.

— Modernisation, functional autonomy down to the individual scientist level, and performance-related reward systems within a merit-based system of promotion are regarded as essential for this purpose.

— The provisions of other commissioned papers concerning the restructuring of ICAR headquarters and on staffing and remuneration.

In ways consistent with the thrust of the Johl Committee, a number of recent initiatives have sought either to supplement existing core funding for research by contract funds, or more radically, to replace part of the core funding with contract funds. Initiatives of this kind : are generally premised on the belief that research becomes more demand-driven, and hence more relevant to the needs of clients, when clients are able to "commission" research. In a number of contexts, additional funds have been established to allow the "clients" of research and extension (usually farmers and ' the organisations representing or working with them) to contract research from the public sector or elsewhere on issues they find important but which are not currently addressed.

These provisions include, for instance, the provisions for contract research made by the government of Rajasthan under the Agriculture Development Project. In such cases, the public-sector research institutions are expected to seize the opportunities offered, since such funds generally permit operational costs to be met adequately and so overcome what was identified above as a major current bottleneck in the public sector.

In a more radical case, the Diversified Agricultural Support Project under consideration by the government of Utter Pradesh is geared towards promoting- diversification and intensification of agriculture j towards market-oriented commodities such as oilseeds and pulses, horticultural crops, livestock and sericulture. Critical of the lack of j infrastructure and operational funds in Uttar Pradesh for research into these commodities, and of the lack of research focus and coordination, the project proposes to allocate over US$23 million to address key production and processing constraints, upgrade research facilities and human resources, and strengthen the research-management system. Proposals for research under the fund should aim to produce time-bound solutions to priority problems using multi-disciplinary, team-based approaches. The fund is open to applications from public-sector and private, commercial organisations, and from NGOs, or from consortia among these.

In reality, public-sector response to contract-funding possibilities has been less than enthusiastic. One reason for this is that performance-assessment criteria in the public sector have traditionally been based on some measure of academic performance, with in some cases recognition for practical achievement such as the release of new varieties. This will have to change as self-earning requirements are imposed in the ICAR, and as the Johl Committee's provisions on, for example, retained earnings are accepted. However, if SAUs do not introduce similar changes, it would remain possible for scientists to be widely promoted on the basis of seniority in return for minimal amounts of research work-on the grounds that operational costs for research provided by the public sector are "unavailable".

Another problem is that many parts of the public sector research system-whether based in the ICAR or the SAUs-are unaccustomed to time-bound, problem-focused research of the kind likely to be financed by grant funds. Research of this type may involve, for instance, the screening of available information to identify what crop varieties already released might be appropriate for specified conditions. Nevertheless, to farmers it represents an important function, and they expect researchers to do it. Performance-assessment and reward systems for researchers therefore need to be modified to encourage them to take up work of this kind. If they do, then this will create a demand which does not currently exist in the public sector for information on the characteristics of available varieties, and so will reinforce the creation of a detailed computerised database.

Several other preconditions also exist for the successful implementation of contract-research schemes of this kind. One is that ideas for research cannot be expected to emanate from low-resource farmers or farmers' organisations unless their inexperience in dealing with government officials has first been overcome. There are by now numerous cases in which NGOs have strengthened farmers' groups to overcome these shortcomings. However, NGOs themselves in many cases have limited technical expertise. In an effort to overcome this, part of the agricultural research fund established by the Ford Foundation in Udaipur has been set aside for independent agricultural experts to work with NGOs and farmers' groups to identify researchable problems and to set out research specifications for them.

Social-science skills have a major role to play in resolving these constraints. These constraints include the following:

— *Enhance the willingness of research staff to identify and respond to clients' needs:* This requirement poses some of the most intractable challenges, including the introduction of merit-based promotion, the redefinition of staff appraisal and performance criteria to reflect client-orientation more strongly, and the rigorous implementation of these. Many of these changes have been introduced in the private, commercial sector in India, and in larger NGOs. It is now for the public sector to follow suit.

— *Fine-tune the financing of research:* Two types of challenge exist here. The first is of a facilitating and incentive-enhancing kind: close monitoring of the implementation of self-earning requirements and a gradual increase in these requirements are desirable, as is a rise in the proportion of research conducted under contract funding. These measures need to be reinforced by full implementation of the Johl Committee recommendations. The implications of these measures will need to be set out' clearly to institute directors to avoid the confusion currently existing in some parts over whether, for instance, contract earnings will lead to a directly corresponding reduction in core funding.

The second challenge is more of a cost-saving and efficiency-enhancing kind. The flow of funds made available to institutes over the financial year clearly needs to be made more even.

Furthermore, some consolidation and rationalisation of research infrastructure is required, particularly in the large number of regional or sub-stations operated by the ICAR centres and the SAUs. Cost savings in this area will contribute towards a further desirable goal: a shift in the balance of resource allocations by ICAR institutes and SAUs to allow more adequate funding for the operational costs of research. The demand for operational funds will rise substantially if, more research is to be done on farmers' fields.

— *Enhance the complementarity between ICAR and SAD research:* Previous efforts in this direction have demonstrated that only limited gains can be expected from formal provisions and over-arching plans. However, gains can be expected from the proposed strengthening of research-information systems. For instance, if scientists are required to use such systems to demonstrate that their research proposals take adequate account of previous or on-going research, then the extent of overlap at the project and programme level will be reduced. The Agricultural Research Information System project initiated by the ICAR during the Eighth Plan for computerisation of ICAR research institutes and SAUs needs to be strengthened in the Ninth Plan. Achievement of this depends largely on the quality of research management-which is discussed below.

— *Allow problem-area focus to influence the structuring of programmes within institutes:* A research institute organised entirely on a discipline- or commodity-basis lends itself to the formulation and implementation of research deriving largely from 'scientists' own preferences. By contrast, one organised largely on the basis of carefully prioritised problem areas encourages disciplines to be drawn into addressing clients' problems. Certain institutes have made radical shifts towards organisation on a problem-priority basis, and provide a useful example to other institutes and programmes.

— *Strengthen project-cycle management:* Major opportunities exist for strengthening the research-project cycle. At the preparation stage, clients need to be identified more clearly, and a careful needs diagnosis conducted. Researchers need to use an adequate computerised database of current and completed research projects to pitch their own proposals in the context of other research. Social scientists can help particularly in ensuring that proposals indicate more clearly than at present the anticipated economic and social impacts of the proposed research. The pro forma for proposals should be modified to accommodate adequate statements of all of the above. As a medium-term goal, it is desirable to require all proposals to be designed in logical framework format, to ensure their internal consistency along with a clear specification of purpose, outputs and activities, and the incorporation of appropriate indicators of achievement at each of these levels.

Procedures for the approval of research projects need to be strengthened. In most cases, ICAR scientific research committees consist only of the full complement of research staff. This has three negative consequences:

— it is difficult to reject proposals in this forum, since to do so would cause the scientists concerned to lose face in the presence of their peers;

— certain disciplines are likely to be under-represented on the SRCs, especially where, as is the case with social scientists, most institutes employ only few, often junior, members of staff; and

— clients are poorly represented. As research becomes more client- and problem-focused, it is important that representatives of the client base should influence the selection of projects.

The implementation of projects can be greatly strengthened. As a minimum, there needs to be continuing consultation with clients as the research progresses. For many types of research there are compelling arguments for experiments to be conducted on farmers' fields and with the strong participation of farmers. The progress of projects-especially the longer ones- against objectives needs to be monitored carefully. At the review stage, lessons for further work need to be drawn out, and feedback from clients needs to be encouraged and acted upon.

— *Adjust the skills balance within institutes:* A higher proportion of research for rainfed farming needs to be conducted off-station and on-farm, that researchers will have to make a substantial effort to elicit farmers' needs, that farmers' risk-averting strategies will have to be understood if the recommendations arising from research are to be relevant to them, and that greater sensitivity is needed in ensuring that the new technologies are environmentally and institutionally sustainable, and that they generate acceptable distributional equity. Social-science skills are needed to meet all of these requirements. Our interviews, already provided strong evidence that social scientists in an ICAR institute have contributed enormously to the successful orientation of research. For instance, close relations between social an technical scientists at the CSSRI over two decades led to recognition that farmers are much more willing to take up soil improvement measures if they include an income-earning component. This led to a wide re-orientation of work towards, for instance, crops capable of tolerating or improving adverse soil conditions which would provide farmers with at least some income, in place of, physical measures.

Three ways need to be pursued for strengthening social-science skills in the ICAR institutes and programmes for rainfed farming research,

— The present allocation of only 7.5% of ICAR staff as social scientists needs to be increased substantially to permit the employment of more (and more senior) social scientists,

— Regular interaction needs to be organised among the social scientists to promote cross-learning from their experiences and to familiarise them with new techniques as they become available. NCAP might play a role in co-ordinating this,

— Biophysical scientists themselves need to develop skills in some areas of concern to social scientists. How this might be done is discussed below.

— *Human resource development.* Strongly client-oriented scientists will wish to be updated with skills, some of which are generic, but all are essential to research for rainfed farming. These include:

— Needs-assessment techniques, including the role of such methods as participatory rural appraisal.

— The preparation of research proposals, including logframe formats if, as recommended, these are adopted by ICAR.

— Management of the research-project cycle, from preparation through implementation to review, including the conduct of participatory, on-farm research and the eliciting of feedback from farmers.

— Understanding of farmers' risk-averting practices and their implications for the design of research and the formulation of recommendations.

— Improved understanding of current neo-liberal concepts concerning the respective rotes of government, commercial and non-profit organisations in technology generation and dissemination for low-resource farmers.

— The preparation and management of research and service contracts with commercial organisations and NGOs; the management of patenting, licensing and royalty arrangements;

— Modes of working in multi-organisation partnership with research and extension agencies within and outside the public sector.

For instance, they have played a key role in developing business strategies in NDRI's work for commercial dairy-processing companies; at CSSRI, they have helped to re-orient research priorities, and in CSWCR&TI they have contributed to economic assessment of aspects of natural-resource management. Elsewhere, however:

— There are few social scientists in ICAR institutes; within this, there is a disproportionately high representation of economists with very few in such disciplines as sociology or anthropology.

— Social scientists do broadly two types of work: providing a "service" function to biophysical scientists by performing basic economic analysis in support of their projects; and conducting farm surveys, often to demonstrate the impact of one or other technology.

— In many institutes, social scientists are a small and relatively young part of the scientific cadre, and often feel a sense of professional isolation.

The curricula used by institutes mandated to provide graduate and postgraduate training for scientists require revision to ensure a uniformly high standard of training in these new areas. For scientists in post, organisations such as NAARM should give in-service training on a regular up-date basis. Prior measures need to be taken to ensure that the faculty of the organisations concerned are familiar with the issues and have adequate teaching and training materials.

Programmes for the exchange of scientists between the ICAR institutes on the one hand, and private commercial organisations and the larger technology-oriented NGOs concerned with rainfed farming on the other, would be of considerable benefit in providing practical experience to reinforce this training. Overall, collaboration among different kinds of institutions in research remains weak outside formal mechanisms such as the AICRPs.

One problem is the fear of the possible financial implications on travel and contingency. Under the prevailing pressure on resources, directors are not willing to spend on scientists from other institutions. There are also fears of credit sharing. The potential benefits arising from savings from enhanced quality of research and reduced duplication of effort are ignored at the level of institute directors, yet have important implications for the system as a whole. There seems to be a lack of understanding of how scientists' time should be budgeted among the departments of the same institute and across institutions. The system should provide for hiring scientists across SAUs and ICAR institutes. This may also resolve to some extent, the non-availability of experts in different disciplines in remote and difficult areas.

Organisation of Extension Services

Extension falls primarily within the mandate of the individual states, and so there are variations in practice from one state to another. Most have used the training-and-visit system of extension, and they are still following T&V's basic framework with some modifications. A review of evaluation studies on the T&V system revealed its impressive gains in irrigated areas, because of the similarity between the agro-ecological conditions where technologies are generated and where they are to be used, and the favourable socio-economic situations and developmental infrastructure for their wider uptake. But there are serious limitations in rainfed areas, where the above conditions do not hold. The states have responded in part to some of the above shortcomings.

One response was to make extension "broad-based": ie, adding advice on land- and water-use, watershed management, agro-forestry and fodder production, in addition to the traditional extension focus on cereals, oilseeds and pulses. One intention behind this modification is that it should permit government departments such as horticulture, animal husbandry and soil conservation to formally delegate responsibility for extension work to the DoA, which generally has the largest network of field staff. However the results of broad-basing have not been encouraging, considering the poor technical background of VEWs and weak linkages existing between different line departments.

Table 2 : Weaknesses of existing extension approaches

Problem area	*Description*
Spread	Technology transfer remains more or less restricted to contact farmers, partly because of faulty selection of contact farmers and the underlying assumptions. Visits become routine as VEW has no new knowledge or skills to offer. VEW and AAO/AO positions in many remote areas remain vacant throughout the year.
Content	Emphasis on field crops; neglect of other crops and enterprises. Low relevance of technical messages to low-resources, risk-prone farming conditions, as they had almost universally been developed from research station experiments. Information delivery as the major role of VEW (which would work well in individual-based technology) and no programme for group-based technologies (such as land-and water-use management) that are important in rainfed areas.

Training	Fortnightly trainings are poorly conducted. SMSs could not be chosen and posted to fit their role. Inadequate subject-matter knowledge of SMSs result in routine sessions and an adverse impact on upgrading and skills of VEWs.
Linkages	Failure of district-level monthly workshops and the Zonal Research and Extension Advisory Committee (ZREAC) meetings held once in a six months to achieve their intended purpose. Poor representation of line departments, and training sessions commonly handled by junior and inexperienced university scientists. Less-than-optimum outcomes, with both sides blaming each other for failure. Linkages at the state level, with wider representation of senior university and state department-of-agriculture officers, bring out serious problem for research-but these are not followed up adequately by either side. Little is done to set research priorities for next year. No specific attempts to acknowledge or collaborate with extension efforts of other agencies.
Operational issues	Total dependence on oral communication. No periodic upgrading of mass-media facilities. Severe shortage of operational funds affecting mobility and other field operations, resulting in inefficient use of human resources.

The need for a group approach in extension began to be accepted later. The Group Farming Programme in Rice Cultivation initiated in Kerala was a notable example. In Rajasthan, the most significant change was the introduction of group-based approaches to extension, in which Kisan Mandate, comprising a group of 20 farmers, were formed in every VEW circle. To facilitate other farmers to meet and discuss with VEWs, Kisan Seva Kendras were opened, whereby the VEW would be available during a specific period during the day of his visit. NGOs also started to play a greater role in technology-transfer activities by the end of the 1970s To increase the involvement of NGOs in efforts to strengthen the research-extension delivery system, a pilot central government scheme, agricultural extension through voluntary organisations, was launched in 1994-95.

The Department of Agriculture in Rajasthan has been experimenting with newer arrangements of providing extension. One major decision in this line was to withdraw the existing extension officials from certain assistant agricultural officer (AAO) circles, and to hand over extension responsibilities there to NGOs. Thus the extension responsibilities at Sivadaspura AAO circle in Jaipur district were handed over to the Social Policy Research Institute, an NGO. Similarly Ras Morarka GT Foundation and Research Centre was given one AAO circle in Jhunjunu district for extension activities. Basically these NGOs are implementing the same programmes of the public system in these areas, with the similar staff positions appointed by the agency. But they are expected to try innovative approaches in implementing the same programmes. Their results would decide the future of such programmes.

Assistance is being provided to some other NGOs also for specific extension activities, such as grain storage. In animal husbandry, a heifer development project has been given to Gayatri Siksha Sadan Sansthan (GSSS) to implement this programme in Saroda block of Udaipur district. In a short period, GSSS succeeded in obtaining critical assistance such as credit and marketing, and provided diagnostic and disease-treatment facilities by establishing and managing a veterinary polyclinic.

Para-Extension Workers

Although public-sector extension services in theory have a strong cadre capable of reaching down to village level, in practice it is extremely difficult to maintain an effective service in many rainfed areas, especially in predominantly tribal districts. Poor communications and transport infrastructure, weak health and educational facilities, and remoteness from district or state capitals: all conspire to make appointments on the lowest rung of the public-extension service to difficult areas appear as "punishment postings". These positions may be filled immediately after a recruitment round, since they provide candidates with an opportunity to gain a foothold in public service. However, as soon as they are appointed they seek transfers to more desirable locations and as estimated, for these reasons, some 40% of the AS positions in Udaipur district, having large pockets of tribal population, are chronically vacant. In many other positions, the rate of staff turnover is so rapid as to render the post ineffective.

Many NGOs in Udaipur district have proposed an alternative scheme in several areas to overcome lack of manpower and other constraints. In brief, their proposal is based on the principles of farmer-led extension, where farmer representatives, elected by farmers themselves, make contact with either ASs in neighbouring areas, or preferably, with officers at a higher level to draw down advice or technologies relevant to local needs. A wide variety of models of farmer-led extension have been tried in countries of Asia, Africa and Latin America, but the most successful have two features in common. One is that farmer-extensionists have to be remunerated if they are to operate on anything other than a very small scale.

The second is that they need initial support to gain confidence to deal with government officials. This second requirement is met in many cases by a programme of training and support provided by NGOs. The former requirement may be met initially by special project funds. In Udaipur, grant applications have been made by NGOs working with farmers' groups and PEWs for the funds for privatising agricultural extension under the government of Rajasthan's World-Bank-supported Agricultural Development Project. Ultimately, there is no reason why they should not be met through a combination of contributions from government, drawing on the funds saved through the persistent AS vacancies, plus contributions from the farmers' groups being served by the PEW.

Apart from meeting the immediate requirements of obtaining advice and technologies from higher levels of the extension service, the PEW approach has a number of other potential advantages:

— Where researchers have taken an interest in participating with farmers in the setting of research priorities, the testing of technologies and the feeding back of farmer' responses into refinement and adaptation of the technology, there PEWs provide a convenient and representative way of incorporating farmers' views into these facets of research.

— Where farmers and supporting NGOs seek to improve local-level seed supply, PEWs can play a lead role in, for instance, establishing contact with local commercial agencies, contacting NGOs, special projects and the like to identify what potentially appropriate varieties exist outside the public-sector seed-supply system. They can also visit NGOs and farmer groups elsewhere to learn from existing approaches to local-level seed supply.

Despite their value in delivering farm-related information and acting as a link with research and extension, PEWs still have a limited base of technical knowledge. They, therefore, cannot be expected to contribute much to the design and implementation of group-based and knowledge-intensive technologies.

MULTI-AGENCY APPROACHES

Many farmers have long been outside the political and economic mainstream, and so are unable to assert their requirements when talking to government officials. Ideally, scientists' training should include skills in both of these areas: in on-farm trials with farmers, and in the use of interpersonal skills to elicit farmers' requirements. Additionally, social-science skills in both the ICAR and SAD system should be strong enough to support and refine these skills. In reality, both of these requirements are unlikely to be met in the short or medium term, and a strategy of multi-agency collaboration with NGOs and farmers' groups has been adopted in some areas-in part to make good these shortages of skills, though it also has wider advantages. For instance, NGOs have skills in group formation and the promotion of joint action which government services are unlikely to develop.

Joint action is necessary in certain agricultural operations, such as pest management, the management of irrigation supplies, and the achievement of the necessary levels of isolation for seed-production plots. It is also necessary for the sustainable management of common-pool resources and their integration with the management of private agricultural land on a micro-watershed basis.

The central premise underlying Manage's training in watershed management is that the technical aspects are less important than the participatory efforts. These are necessary to get joint action on common-pool resources organised, and to be able to identify and resolve potential conflicts over common-pool resource management-before on-the-ground efforts to improve resource management are made.

Manage's approach makes a clear distinction between two aspects of extension which have been conflated in conventional approaches: skill development and information dissemination. Manage argues that the latter is now largely transmitted by mass media than by extensionist-farmer contact. The former, it argues, is best spread by farmer-to-farmer contact, with input from government technical agencies, which limit themselves to expanding the basket of choices available to farmers.

Manage's concept of "distance learning" embraces both the use of multimedia at local level, and the spread of skills via farmer-to-farmer contact. Manage's view is that, whilst some agricultural research in India has been problem-driven and so has at least put additional choices "on the shelf, virtually all extension has been top-down, and driven by targets. The last decade also saw the emergence of numerous organisations in the field of transfer of technology. This includes, other than the public sector, organisations in the private industry, NGOs, farmers' associations, cooperatives and private consultants.

ICAR and Extension

Although extension is predominantly a state responsibility, central government, in the form of ICAR, retains responsibility for "frontline" extension, and the ICAR continues to make major financial contributions to the KVKs, which have responsibility for demonstrating and providing training in new techniques. From the beginning of the Eighth Plan, KVKs were mandated to:

— Collaborate with the SMSs of the SAUs, scientists at the ICAR and the SAUs, and with state extension personnel in "on-farm" testing, refinement and documentation of technologies for developing region-specific sustainable land-use systems.

— Organise long-term vocational training courses in agriculture and allied areas for rural youth, with emphasis on "learning by doing" to generate self-employment opportunities.

— Organise front-line demonstrations on various crops to generate production data and feedback information.

— Organise regular training to update state extension personnel in advances in agricultural research.

In brief, KVKs are to provide a key facilitating role in the refinement of technologies to specific conditions, by acting as a two-way link between research and farmers. The staffing provided by the ICAR includes the disciplines of agricultural extension, agronomy, plant breeding, horticulture, veterinary and animal sciences, home science and plant protection.

CURRENT ISSUES

The ICAR's involvement in extension activities started with the National Demonstration Scheme initiated in 1966. Since then, innovative programmes have included the Operational Research Project, Krishi Vigyan Kendra, Lab-to-Land Programme, All-India Co-ordinated Research Programme for the Upliftment of Scheduled Castes, and the Tribal Area Research Project. From the Eighth Five Year Plan all these programmes were merged into KVKs. All have been 100%-funded by ICAR since their inception.

ICAR now supports 261 Krishi Vigyan Kendras, of which 27 are in ICAR institutes, 139 in SAUs, 86 in NGOs, and the remaining nine in central universities, agricultural colleges, etc. Another 58 KVKs are to be established by upgrading existing Farmers Training Centres through NATP funding. On average, ICAR spends Rs 32 crore yearly on the KVKs (Eighth Plan). In addition, ICAR spends more than Rs 10 crore on technology assessment and refinement through institute-village linkage programmes in selected centres. The proposal was first implemented in rabi 1995-96 on a pilot basis, and was expected to cover three seasons at 42 selected centres, with subsequent up-scaling in the National Agricultural Technology Project.

The increasing costs of establishing and maintaining Krishi Vigyan Kendras became an issue of serious concern by the late 1970s ICAR Plan funding for extension has increased from 3% of all funding under the Sixth Plan to over 12 % in the Eighth Plan. At present, the Council's decision is to fund existing KVKs and those to be started from 1992-93 for only 10 years. ICAR's decision to phase out its funding gradually has invited criticism from the existing KVKs and the political establishment.

Several committees have looked into the functioning of the KVKs. The Evaluation Committee on KVK, the Joint FAO-ICAR Committee on KVK, the Quinquennial Review Committee on KVK and the Committee on HRD through KVKs and TTCs are some. The review committee on the ICAR extension system has also looked at the functioning of the KVKs. One conclusion that can be drawn from all these reports is that KVKs have been an institutional innovation and have contributed towards training farmers and rural youth. Visits made by the team to three KVKs based at NGOs Vidya Bhavan Society, Youth for Action, and Agricultural Development Trust also support this.

However there are weak KVKs, too. Following a visit in 1994, Van den Ban noted that many KVKs are under-resourced and have inexperienced staff. The World Bank noted that many KVK training courses were under-subscribed, raising doubts about their relevance. There is a need for an objective and scientific evaluation of all KVKs so that a case-by-case assessment will guide the type and level of any further support.

KVKs are mandated to cover hands-on training for farmers, farm women, rural youth and village extension workers on different aspects of agriculture, animal husbandry and rural enterprises. There is no other comparable organisation existing at the district level, and this itself is a very strong argument for the KVKs to continue and expand. Some argue that this is basically a function of the state department of agriculture, and that there is duplication between ICAR extension through KVKs and the extension activities of line departments. Our experience is that the state departments of agriculture, following the T&V pattern, have been concentrating on information transfer and implementation of other schemes.

The Farmers Training Centres have generally performed poorly, and the present proposal is to upgrade them to KVKs. As the technological options in rainfed agriculture are going to be knowledge-based, resource-sharing and complex, farmers and the facilitators need to be provided with hands-on training. Ideally, these are the functions that the central Ministry of Agriculture or respective state departments of agriculture ought to perform.

The Surjewala Committee recommended 100% funding to KVKs in the Ninth Plan, a one-time project fund for generation of income, continued funding of 75% for the next 5 years, and 50% funding forever. If the ICAR has to manage these KVKs, it needs additional support, otherwise its allocation for other functions such as research and education would be affected adversely. The issue is, who would support the KVKs.

KVKs should be in a position to generate a part of their resources from the sale of planting materials and other produce from their farms. Training programmes can also be charged for to some extent. KVKs are also encouraged to seek long-term funding relationship with local constituencies, such as NGOs, and with national and international organisations. By these means, KVKs would be able to meet at least their operational expenditures other than salaries. Salaries would have to be provided by ICAR through a grant or special allocation from the Ministry of Agriculture or state department of agriculture, or even from the Ministry of Rural Development. The financial allocation to the different departments within Ministry of Agriculture is given in Table 3.

Table 3: Financial allocation to Ministry of Agriculture departments, 1995-96

Department	*Rs. (crores)*	*Percentage*
Department of Agriculture and Cooperation	1,490	69.5
Department of Agricultural Research		

and Education	310	14.5
Department of Animal Husbandry and Development	344	16.0
Total (Ministry of Agriculture)	2,144	100.0

While DAC gets about 70% of the total MoA allocation, the ICAR share is less than 15%. Out of this, ICAR spends more than 12% on the KVKs, which are essentially an extension function. In the KVKs, 1 ICAR has institutionalised and perfected an innovative set-up for transfer of technology, and the DAC may now come forward to support it. Rs 32 crore of ICAR's expenditure on KVKs would presently come to less than 3% of the DAC budget.

The current ICAR funding for KVKs is under review and is likely to be reduced. The goal set in the Quinquennial Review of KVKs in 1994 of establishing one KVK per district by 2002 is therefore unlikely to be achieved. A far-sighted move on the part of ICAR was to locate a number of KVKs at NGOs, in the belief that NGOs' ability to interact with farming communities, especially in the more difficult rainfed areas, was key to the identification of farmers' needs. However, the performance of the NGOs concerned is mixed. Some, such as the Vidya Bhawan Society in Udaipur have provided extensive support to their KVKs. Others, however, suffer from the common NGO characteristic of limited technical orientation, and so have found it difficult to develop a strong working relationship with-their KVK.

The World Bank report noted that many of the long-standing programmes that had been allocated to KVKs, including the Lab-to-Land programme, are "paternalistic in conception and execution", taking scientific endeavour well beyond its accepted mandate. In the flush of enthusiasm for the Zonal Research Stations then being established, it also argued that all ICAR technology-transfer projects be wound up, or continued by the states where they found it desirable, under the mandate of the ZRS. As mentioned above, others have questioned whether there is duplication of effort between the KVKs and Farmers Training Centres of state governments.

A number of changes have recently been made in ICAR's relations with the KVKs, most of them relating to funding. By March 1997, it was envisaged that ICAR support would be withdrawn from the 37% of oldest KVKs, with withdrawal from a further 33% by 2002 and the remaining 30% by 2005. The chief training organisers of KVKs have always been given flexibility in identifying appropriate programmes and sources of funding beyond that provided by ICAR for the areas in which they operate. One advantage of the new funding policy is that this flexibility is reinforced: CTOs are being positively encouraged to seek long-term funding relationships with local constituencies, such as NGOs, and with international organisations.

Other changes have been promoted by the ICAR Division of Agricultural Extension to encourage a participatory interaction directly between selected ICAR institutes, SAUs, KVKs, Regional/Zonal Research Stations and villages in the Institution-Village Linkage Programme (IVLP). This proposal, first piloted in rabi 1995-96, was expected to cover 42,000 farmers over three seasons at 42 selected centres, with subsequent up-scaling by the National Agriculture Technology Project to some 200,000 farmers. Basically, the IVLP argues that technology integration- process or set of activities that must happen if a new information can be put to use in the farm production system- is a neglected function.

In the IVLP view, this neglect accounts for the widespread failure of technologies to be taken up, especially in complex, diverse and risk-prone rainfed areas. The IVLP recommends problem-focused, systems-oriented, multi-disciplinary approaches capable of generating "baskets" of technologies that are profitable, highly sustainable, low-risk and equitable in their impact among men and women. The socio-economic impact of such technologies will be assessed, as will their capacity for scaling up to other areas. Relatively underdeveloped villages representative of a district will be selected, and participatory rural appraisal will be used to diagnose technology requirements. Teams comprising four or five of the main disciplines will be selected and trained in the necessary client-oriented, participatory techniques of identifying needs and screening, testing and adapting technologies.

We appreciate the need for technology assessment and refinement, but have doubts on the methods to address this through the IVLP programme. Developing appropriate technologies are the responsibilities of all concerned with technology development, and should not be the task only of a team of scientists at the institute level. Agro-ecosystem analysis reports would become useful only if there is a willingness to redesign research based on their findings.

Other than studying farmers' socio-economic conditions, farming practices, etc, which are important, social scientists should enhance their understanding of the local economy, existing production networks and system interconnections in their area. To design appropriate models of technological interventions, social scientists may have to increasingly adopt action-research strategies. The contribution of social scientists in the ICAR/SAD system in this line is at present very limited.

REFERENCES

Alsop, R., G. Khandelwal, E. H. Gilbert, J. Farrington, *Human Capital Dimension of Collaboration among Government, NGOs and Farm Families: Comparative Advantage*, Complications and Observations from an Indian Case, Agriculture and Human Values 13 (12), Spring 1996.

7

Agricultural Biotechnology Research

Around 35 years ago, agricultural production in India got a major boost with the introduction of dwarf varieties of wheat and rice. The introduction of these varieties led to a dramatic increase in the yields of the two crops. Some productivity enhancement came through the use of hybrids in corn, sorghum and millet although the area under cultivation of these crops has steadily decreased in the last decade. In the last 10 years the yields of rice and wheat have also plateaued out.

The productive agricultural areas in the North, due to continuous rice–wheat cultivation are encountering serious problems of sub-soil water depletion, deficiency of micronutrients in the soil and increase in the use of pesticides, fungicides and herbicides to control pests, pathogens and persistent weeds. Agricultural production is becoming more and more dependent on agrochemicals, thereby increasing input costs and causing significant damage to the environment and human health. Both farmers and consumers are at the receiving end –farmers by exposure to agrochemicals and consumers due to residues of agrochemicals in the consumed food.

While there is self-sufficiency in cereal grains at present, the yields and productivity of dryland crops, mostly grain legumes and oilseeds remain low and no major breakthroughs in productivity enhancement and yield stabilisation have been achieved. Currently, India is importing both grain legumes and edible oils to meet internal demands. India's population is expected to reach approximately 1.5 billion by 2050. It is estimated that around 300 million of India's population is suffering from malnutrition.

Due to socio-economic reasons, women and children are more at the receiving end in terms of micronutrient deficiency and low calorie intake. Nutritional security for everyone would require more extensive availability of grain legumes, edible oils, fruits and vegetables, milk and poultry products. The challenges of malnutrition, enhanced productivity and crop diversification can be met by better resource management management and by breeding more productive, more nutritious and at the same time less resource input demanding crops.

Crop biotechnology, which broadly includes areas of development of transgenic crops, structural and functional genomics and marker-assisted breeding could provide us with the vital breakthroughs to achieve improvements in both quality and quantity in a sustainable manner.

With the advent of techniques of genetic engineering in the early seventies, the natural barrier to gene exchange has been removed. Sequences from varied sources like bacteria, viruses and eukaryotic systems can be transferred to plants to develop transgenic crop varieties. Achievements, to date, in plant biotechnology have surpassed all previous expectations and with the development of high throughput instruments, the future is even more promising.

RESEARCH IN INDIA

Recognising the importance and potential of biotechnology, the Government of India set up the National Biotechnology Board in 1982 which became a full-fledged department in 1986 under the Ministry of Science and Technology. The plan-wise budget of DBT from 1992 onwards in the area of crop biotechnology is given in Table 1.

Table 1. Plan expenditure by DBT in crop biotechnology.
Amount (Rs in lakhs)

Scheme	*8th Plan 1992–1997 (Actual)*	*9th Plan 1997–2002 (Actual)*	*10th Plan 2002–2007*	*2002–2003*
Crop biotechnology	1916.00	2897.00	7500.00	1400.00

To provide research and human resource development in crop biotechnology, DBT supported the establishment of Centres for Plant Molecular Biology (CPMB) in 1990. A total of six such centres were set up initially at various Universities/Institutions namely, Jawaharlal Nehru University (New Delhi), Madurai Kamaraj University (Madurai), Tamil Nadu Agricultural University (Coimbatore), Osmania University (Hyderabad), National Botanical Research Institute (Lucknow) and Bose Institute (Kolkata).

A seventh centre was established at the University of Delhi South Campus in 1997. A total financial outlay of Rs 45.98 crores was made available for the development and research activities of CPMBs. Most of the research on crop biotechnology currently underway in India is being supported by the Government and the DBT has been the main agency amongst government departments to support research endeavours in this area.

The DBT has supported over the last 12 years a large number of research projects which deal with the development of *in vitro* regeneration and genetic transformation protocols of important crop species grown in India, the development of transgenics with genes of agronomic importance and several projects dealing with basic research in plant molecular biology.

To bring investigators with expertise in different facets of biology together to address defined research goals, a number of multi-institutional projects have been launched which include development of transgenics for resistance to geminiviruses in cotton, mungbean and tomato, resistance to rice tungro disease, resistance to bollworms in cotton, development of nutritionally improved potato with a balanced amino acid composition using a heterologous protein from *Amaranthus* and development of molecular methods for heterosis breeding.

Another set of projects include tagging of quality traits in rice and mustard, resistance to karnal bunt and rust in wheat and an extensive research network on rice biotechnology, which includes work on genomics, mapping, improvement in nutritional quality and resistance to biotic and abiotic stresses. Another multi-institutional project on tagging of quality traits in wheat by molecular markers has given some interesting leads. A number of markers linked to grain protein content, seed size, bread-making quality, etc. have been identified. To further strengthen research in the area of crop biotechnology, a new institute, National Centre for Plant Genome Research (NCPGR) has been established in New Delhi with a mandate to strengthen plant biotechnology research in India.

Rice Production in India

Rice production in India holds the key to the food security of the country. After detailed consultations with experts at the National and International level, DBT and the Indian Council of Agricultural Research (ICAR) jointly took the initiative to participate in the international rice genome-sequencing programme (IRGSP) in June, 2000. Two laboratories in India, one at the University of Delhi South Campus and the other at the Indian Agricultural Research Institute (IARI) in New Delhi are jointly sequencing part of chromosome II. This region carries genes for a number of biotic and abiotic stresses and genes encoding some agronomically important traits.

Around 15.58 Mb of DNA has already been sequenced till December 2002. A high quality draft sequence of rice genome by IRGSP has been declared completed by the Japanese Prime Minister Junichiro Koizumi on 18 December 2002 in Tokyo. In order to take advantage of sequence information in rice, a workshop was organised in May 2002 to identify areas which should be funded for functional genomics of rice. It was decided to develop research programmes in the areas of:

(i) salinity and drought stress,

(ii) biotic stresses,

(iii) mapping of loci contributing to high yield.

The project on genome sequencing will also allow rapid progress in the characterisation of allelic diversity available in *Oryza sativa* and related species and genera for resistance to diseases and pests of rice.

Future Developments

In the 10th Five Year Plan (2002–2007) the following areas will be given high priority:

— Rice genomics with particular emphasis on the study of genetic variability for achieving both yield increase and stabilisation in the crop.

— Molecular maps in rice, wheat and grain legume crops. To date, a molecular map with reasonable density (around 1000 AFLP markers) has been developed only for mustard (*Brassica juncea*). In wheat, tagging of some interesting traits has been accomplished. The work on cereal crops will continue and mapping work will be initiated in grain legume crops. Molecular maps will be used for mapping of QTLs for abiotic stresses, plant architecture, and yield and mapping of loci conferring resistance to pests and pathogens.

— Transgenics developed in various crops through DBT funding will be tested under appropriate containment conditions and if found useful will be subjected to nutritional and toxicological tests to determine their suitability for cultivation in the farmer's field.

— Research on development of pollination control methods for the production of hybrid seed will be further strengthened. Fixation of heterosis through apomixis will be another area of research.

— Particular emphasis will be given on an improvement in the nutritional status of some of our major crops. To begin with, the three target crops will be rice, wheat and potato. High-yielding rice varieties grown in India will be either transformed with constructs that would code for high β carotene content or by back-crossing these varieties to 'Golden rice' transgenics already developed in Switzerland. Improving the iron content in rice grains will be another thrust area. In potato, nutritionally enhanced transgenic Indian varieties of potato carrying the *ama1* gene are already under advanced field trials. If found suitable, these will be released for cultivation.

— Transgenic approaches to senescence retardation in fruits will be a major thrust area. It is important to reduce post-harvest losses in fruits and vegetables as under tropical and sub-tropical conditions senescence is much more rapid and energy costs of cold-storage are very high.

Safety of Transgenic Material

Safety of transgenic material is a major concern. A major contribution of DBT has been to set up, in collaboration with other government departments, a framework for the evaluation and eventual clearance of transgenic crops for field cultivation. The release of transgenic crops is governed by the Indian Environment Protection Act (EPA) – 1986 which came into force from May 1986. The act provides a framework for the protection and improvement of environment.

Later, the rules and regulations for the manufacture, use, import, export and storage of hazardous microorganisms, genetically engineered organisms or cells were notified under EPA 1986 on 5 December 1989. A mechanism based on interaction between committees and different departments of Government of India has been set up. Such materials will have to meet with the approval of the following Committees: Institutional Biosafety Committee (IBSC), Review Committee on Genetic Manipulation (RCGM) and Genetic Engineering Approval Committee (GEAC).

Under the EPA 1986, the GEAC examines from the viewpoint of environmental safety and issues clearance or the release of genetically engineered organisms/ transgenic crops and products into the environment. The GEAC also grants permits to conduct experimental and large-scale field trials which are beyond the limit of 20 acres. In case of transgenic crops, applicants are required to seek clearance from the Ministry of Agriculture, which will accord the final clearance of transgenic crop varieties for agricultural activities.

Bt cotton was released for commercial cultivation in March 2002, following the biosafety procedures laid down by the Government of India. Three cotton hybrids have been granted permission for field sowing and are currently in the fields in six states namely, Maharashtra, Gujarat, Madhya Pradesh, Andhra Pradesh, Karnataka and Tamil Nadu. Transgenic mustard hybrids are also in advanced field trials and if found suitable, may be released for general cultivation.

NATIONAL RESEARCH CAPACITY IN DEVELOPING COUNTRIES

Research capacity is the portfolio of resources – physical, human, and financial –available in a research system for performing and utilising research. Research capacity is not limited to undertaking research projects alone, but includes engagement with a broader innovation system, including specifying, accessing, interpreting and applying research. For agricultural biotechnology, research capacity also involves a spectrum of key elements and activities, including:

— Defining objectives and priorities for agricultural biotechnology;
— Developing and implementing a clear policy for (agricultural) biotechnology, and biosafety regulations;
— Developing R&D management capacity;
— Transfer of technologies, knowledge and skills to the private sector;
— Promoting international collaboration and technology transfer.

How such research capacity creates outcomes requires a broader discussion of approaches and methods. With the involvement of stakeholders comes new research objectives for public sector institutions. The tendency now is for research systems to develop capacity that is flexible, rigorous and which actively involve a wider range of stakeholders. This includes approaches taking into account social and economic considerations, and other approaches allowing public research systems to address environmental and biodiversity concerns, resource degradation, public health issues such as HIV/AIDS, and other critical problems affecting agriculture.

Challenges in Agricultural Biotechnology

Capacity building in agricultural biotechnology faces unique challenges when compared to other technological revolutions, including the Green Revolution in the 1950s and 1960s. These challenges, either directly or indirectly, affects capacity building, retention of personnel, and the balance between public and private sector capabilities. These challenges include:

Development costs

Whether a country is an innovator in agricultural biotechnology, or whether it adapts technology developed elsewhere, getting involved will entail new investments in research infrastructure, tax breaks or subsidies for pioneering companies, training and management development for research institute personnel.

Competition with conventional research programmes

There is a concern that resources devoted to germplasm conservation and enhancement as well as conventional plant breeding activities may be re-directed towards biotechnology research.

Corporate control and distributional implications

Private sector companies have primarily researched, developed and marketed biotechnology products. This is a departure from the "Green Revolution" technologies of the 1950s and 1960s, such as semi-dwarf varieties of wheat, developed primarily by the public sector. Initially, the private sector held exclusive ownership of biotechnology innovations. This opened the possibility for private sector companies to exercise market power in the seed markets.

Economist's wariness about this issue is a result of concentrated market structures being associated with above normal prices and extraction of rents from producers who buy inputs. In the specific case of seed markets, some authors propose that seed prices are already higher than normal, and that the long-term goals of enhancing germplasm and improving genetic diversity have suffered with increases in market concentration.

Intellectual property issues

An expanding body of literature shows that increasingly both private firms and public institutions claim property rights in agricultural research and biotechnology. Property rights instruments include patent rights; plant variety rights; and contractual rights arising from material transfer agreements (MTAs). The strengthening and enforcement of intellectual property rights (IPRs) throughout the world have dramatically influenced the processes for research collaboration and international transfer of new technologies.

While there is no conclusive evidence whether stronger IPRs enhance or impede the availability and diffusion of new technology to developing countries, it is important that research institutes carefully consider their IP management strategies and "freedom to operate" prior to embarking on new R&D projects. Given the general situation of "weak" protection of biotechnology inventions in most developing countries, and the fact that patents for key enabling biotechnologies are primarily in force in highly advanced countries, problems of intellectual property infringement may not be too serious. This could of course change in the future.

Biosafety regulatory issues

Although there has not been one documented case, so far, of any environmental or human health damage caused by GMOs, there are some aspects of the risk profile that are not known, simply because the accumulated knowledge derived from the commercial-scale release of GMOs is still limited. This lack of familiarity has resulted, in a large number of countries, in stringent biosafety regulations and sluggish decision-making, thus adopting a precautionary approach to the diffusion of GMOs. It is important to point out that research can enhance familiarity with biotechnologies and contribute to informed biosafety decision making through risk assessment studies.

Consumer acceptability

Consumers in Europe, to a lesser degree the United States and some countries in Asia, have not been supportive of current biotechnology products. There is an increasing body of literature documenting this phenomenon; however, the important factor is that consumer groups in Europe and Japan have been very vocal about their opposition to biotechnology and have managed to exercise political pressure on governments to stop the development of biotechnology products.

According to Paarlberg, the consequence of this political pressure is that developing countries may have an even smaller probability of accessing biotechnologies suited to answer their problems. This is because developing countries are afraid that Europe and Japan will shun imports from countries using biotechnologies. Furthermore, Paarlberg argues that increased European Union regulations on labelling and identity preservation will further discourage the creation and diffusion of genetically modified crops in poor countries.

Dependence and un-sustainability

Governments need policies and incentive mechanisms to encourage private-sector investment and participation in agricultural biotechnology. Public-and private-sector research should be consciously complementary and not competitive. The policy framework should not only promote the safe use of biotechnology but also ensure that policies are not a deterrent to investment by the private sector and to collaboration with global science.

Each of these actions would help countries to move away from dependence on donor or external funding, and thus make their research more sustainable. This problem is not exclusive to biotechnology, yet the initial costs of developing these technologies may foster this type of problem

Biotechnology Capacity in Developing Countries

Recent reviews of agricultural biotechnology research capacity in different geographic regions show a mixed picture, very much reflecting the overall state of agricultural R&D investments. Reviews on Asia and Latin America confirm a steady development of agricultural biotechnology R&D particularly in relatively large developing countries with well established infrastructure for research: Argentina, Brazil, Mexico and Chile in Latin America; and, China, India, Indonesia, Malaysia, The Philippines and Thailand in Asia.

In these countries, human and financial resources allocated to biotechnology R&D are relatively high, experience with GMO testing and commercialisation is growing, and government support programs and policies actively encourage R&D. It is no surprise that research institutes in the countries mentioned above are also actively involved in bilateral and international collaborative research programs in agricultural biotechnology.

The situation in sub-Sahara Africa stands in sharp contrast with that found in Asia and Latin America. Surveys by ISNAR and IITA and Alhassan indicate that although biotechnology applications are increasingly incorporated in agricultural research programs, they primarily involve applications of cell biology and disease diagnostics, while more advanced applications are found only in a handful of institutes, most notably in South Africa.

Furthermore, research efforts are scattered over a wide range of products and institutes without critical mass, and most often heavily dependent on donor funding. Advancements in agricultural biotechnology are severely constrained by a lack of funds, skilled human resources, and equipment. Government priorities and policies to support agricultural research in general, and agricultural biotechnology in particular, are lacking or not being implemented.

Measuring Research Capacity

Before looking at outputs however, it is first important to understand the investments made in research capacity and infrastructure. To accomplish this, a study was initiated for developing countries that included a series of indicators. To achieve consistency between countries, the survey instruments used in our six partner country case studies were designed to include a common typology of research institutes and core variable definitions. The research institutes included in this study were:

— Private commercial firms (input companies, farm sector and food processing companies)

— Private non-commercial organisations (foundations and non-governmental organisations)

— Public sector (national institutes for agricultural research and universities)

Presently, there are few outputs commercialised by the public sector from investments in plant biotechnology at this point, except for China. However, this may change for some countries included in the survey, especially if the cost of plant biotechnology research decreases over time, and public sector research systems in developing countries gain the skills and infrastructure necessary to produce appropriate biotechnology outputs, as wells as develop dissemination channels for their products.

In these studies, research staff includes individuals who hold a research position with at least a BS degree or its equivalent. Research staff included personnel in Management positions, such as directors and heads of research programs. Personnel data reported in the survey in both head count and full-time equivalent (FTE) researcher year terms and adjustments were to account for part-time commitments.

Research expenditures include personnel, operational and capital expenses. Research expenditures have been deflated to account for inflation. Real expenditures have been converted to an internationally comparable standard by adjusting for differences in Purchasing Power Parity (PPP) Index with base year of 1985. The expenditures adjusted to PPP dollars will be used to make inter-country comparisons of expenditure levels.

Research and Development in agricultural biotechnology

During the period of analysis, total expenditures in agricultural plant biotechnology research measured in real 1985 US$ Purchasing Power Parity totalled 508 million dollars (table 2). Investments in plant biotechnology research increased in China, Indonesia and Mexico. Expenditures for Kenya, Zimbabwe and Colombia remained relatively static during the period of analysis. It is important to reiterate that investments in Kenya and Zimbabwe fluctuated significantly over the period of the survey as a result of donor funding during the period.

The survey data pertaining to the structure, organisation, human resources, expenditures, and financing of biotechnology research are analysed below. The evolution of the agricultural biotechnology research in the six selected countries is analysed in relation to structure and organisation, human resources, expenditures and sources of financing.

Structure, organisation and sources of funding

Public-sector organisations primarily implement

agricultural plant biotechnology research in the six countries included in the survey. Although the participation of the private sector had a low during the period of the survey, this sector had higher annual growth than did the public universities. Moreover, universities showed higher fluctuations and/or declines in research investments in these countries.

This can probably be explained by the cyclical nature of donor funding and economic recession. For example, the departments of biochemistry and crops sciences of the University of Nairobi and the University of Zimbabwe received a substantial lump-sum donation in early 1990s that distorted their research expenditure figures during the period of analysis.

Table 2. Total expenditures for plant biotechnology research in selected countries

Country	*Years*	*# Of Institutes Surveyed*	*Total expenditures for survey period (Million 1985 US$ PPP)*
China	1986-1999	22	158
Colombia	1985-1997	10	15
Indonesia	1989-1997	8	116
Kenya	1989-1996	6	11
Mexico	1985-1997	14	193
Zimbabwe	1998-1998	6	15
Total			508

In the case of Mexico, government funds many universities, so the 1995 recession could be a factor affecting public universities' decrease in allocation of research investments from 1985 to 1997. These results are in contrast with results from other research efforts elsewhere. Byerlee and Fischer indicate that the majority of investments are conducted by the private sector internationally. However, their results are directly influenced by the amount of research done by multinational corporations, particularly in developed countries.

Colombia is an interesting case where coffee producer levies were used to finance research in coffee biotechnology in the mid-1980s. Coffee levies represented up to 59% of plant biotechnology investments in Colombia during the mid-1980s. Public research institutes showed not only the highest share of financial resources but also the highest annual growth rate (Kenya 9%, Indonesia 30%, and Zimbabwe 70%). Furthermore, high annual growth rates were accompanied by a concentration of financial resources on few public research institutes. KARI in Kenya accounted for 70% of total expenditures in 1996, BRI in Zimbabwe 80% in 1998, RIFC, RDCB and IUC in Indonesia 70% in 1997, and CINVESTAV-I, CICY and IBT-UNAM 55% in 1997.

The composition of Colombia's biotechnology research has changed in 1997 where more public institutes and universities, and private non-commercial institutions have increased investments in agricultural plant biotechnology research. On the other hand resources in China are concentrated around the research institutes, although funding at the university level has increased significantly.

The donor share in total expenditures has been considerable at Kenya and Zimbabwe. Donor contributions there represent an average 58% in Kenya and 50% in Zimbabwe of total expenditures on agricultural biotechnology. Public research institutes in these two countries were mainly funded by donor contributions, which were concentrated to the main research institutions. KARI in Kenya accounted for almost 85% of total donor support in 1996 and BRI in Zimbabwe almost 90% in 1998.

Nonetheless, donor contributions to the development of agricultural biotechnology research programs will probably change significantly depending on pledges of support by foreign governments and multilateral development agencies but also on the socio-political climate in the country. It is important to point out that the viability or maintenance of these levels of funding will be compromised in the medium term if there is no effort to get some funding from local sources. Otherwise, the efforts made for the development of agricultural biotechnology could be endangered in both countries.

Biotechnology activities done by private non-commercial organisations are funded by taxes on their commercialised commodities and sales of products. In China, Mexico and Indonesia the government share in total expenditures has been considerable for public research institutes and the biotechnology programs at the universities. Government contribution represents about 88% in China, 61% in Mexico and 83% in Indonesia of total expenditures on agricultural biotechnology.

Donors share, not as significant as in Kenya, accounted for almost 10% in China, 16% of Colombia, 32% in Mexico and almost 7% in Indonesia. It is interesting to note that some public research institutes and universities are funding their biotechnology research activities from non-traditional sources of funding, such as sales of products and services and contractual arrangements. Although these sources of funding are still minimal, they have increased during the period of analysis.

As expected, contracts largely fund biotechnology activities carried out by private non-commercial organisations, while sake if their products finance private commercial organisations. The limited funding (or non-existent in the case of Kenya and Zimbabwe) from nontraditional sources (contracts, sales of services and products) of the public research institutes and universities indicates a limited relation between those public entities and the private sector. Donors' contributions are important sources of funding in Kenya, Mexico and Zimbabwe. A closer look at how donor funds are distributed reveals that most donor contributions were oriented to infrastructure, consultants, and operations.

In the period of analysis, these categories accounted for almost 95% of donor funds. It is interesting to note that the share of donor contributions going to operations and consultants decreased during the period, while infrastructure increased. Sharp increases in the number of researchers and the need to modernise laboratories may explain this fact. On the other hand, it is surprising that donors' share of training is almost nil in Kenya and Mexico. However, in Zimbabwe donor's share to training was almost 20% indicating an interest of building human capacity in biotechnology.

Human resources

In Kenya and Indonesia, the number of researchers more than doubled. In China, Mexico and Zimbabwe researcher numbers quadrupled. In Colombia, the total number of research professionals increased more than 10-fold, but most of the increase in research personnel occurs at the M.Sc. and B.Sc. level. This may be explained by the emphasis on ornamental and floriculture efforts directed particularly to tissue culture. The number of Ph.D. holders increased significantly in all countries.

For example, in Kenya, Indonesia and Zimbabwe the number of Ph.D. holders at least doubled, whereas in Mexico and Colombia the number of Ph.D. increased ten fold. China is a very interesting case where the number of Ph.D. holders increased dramatically from 5 in 1986 to 203 in 1999. These groups of professionals form the basis for future agricultural biotechnology. Although the number of scientists working in biotechnology has increased, they are, nonetheless, concentrated in a few public research organisations.

Around 45% of the researchers are located in KARI (Kenya), 60% in only four research Mexican organisations:

CINVESTAV-I, CICY, IBT-UNAM, and the National Research Institute on Forestry, Plants and Livestock (INIFAP), 60% in only three Indonesian research organisations: RIFCB, RDCB and IUC, and 70% in only three Zimbabwean research organisations: DRSS, BRI and UZ. China is a counterexample where scientists are distributed among a large number of institutes and universities. However, this may be a consequence of the sheer size of the Chinese agricultural research system.

Investment expenditures

Because the rate of growth of the number of researchers is higher than the rate of growth of expenditures, this has led to significant declines in expenditures per researcher. The only exceptions are China with a relatively constant level of real expenditures and Indonesia where there was a small increase level of expenditure per researcher. This trend could change depending on the behaviour of multilateral and bilateral investment institutions and the general state of the countries economies.

It is worthy to highlight that expenditures per researcher in Mexico and Indonesia were relatively higher than those of the smaller countries in the sample (Colombia, Kenya and Zimbabwe). These figures imply that Mexican and Indonesian researchers have more resources and higher probabilities to generate biotechnology research outputs and may be an additional signal from their respective governments about their commitment to biotechnologies in general.

One of the most widely used measures of measuring national commitment to a research agenda is the research intensity ratio. The agricultural biotechnology research intensity ratio is measured by dividing investments in agricultural biotechnology research relative to Agricultural Gross Domestic Product (AG GDP). Even though the agricultural research intensity ratio has grown annually, the percentage of agricultural biotechnology research expenditure in relation to Agricultural GDP is quite minimal, on average 0.002% in China, 0.007% in Colombia, 0.09% in Zimbabwe, 0.04% in Mexico, 0.025% in Kenya and 0.014% in Indonesia.

During the same period the percentage of agricultural biotechnology research expenditures to total agricultural research expenditures was in average 0.55% for China, 1.2% for Colombia, 1.5% for Kenya, 6.4% for Mexico, 6.8% for Indonesia and 4.1% for Zimbabwe. On average, Indonesia and Mexico invested more than the small countries in the sample. There is no set rule for how much of the agricultural research budget should be allocated to biotechnology. The World Bank and SPAAR have suggested a 2% of agricultural GDP as a target for finance of agricultural research by developing countries.

However, there are no guidelines for components within agricultural research. But for comparison, the Consultative Group for International Agricultural Research (CGIAR) spends about 8% of its budget to biotechnology research in 1997, and the United States allocated 13% of its agricultural research expenditures to biotechnology in 1992. In addition, the amount of investments in agricultural biotechnology may be related to the relative strength of the agricultural research system in general. Byerlee and Fischer in a rough classification of National Agricultural Research Systems (NARS) based on the plant breeding and molecular biology capacity, indicate that "stronger" NARS may invest more in agricultural biotechnology than "weaker" systems.

The authors estimate that the stronger NARS (Brazil, China, Mexico, India, and South Africa) indeed invest 5-10% of their budgets in agricultural biotechnology applications. Huang, Rozelle, Pray and Wang highlight the situation in China. According to these authors in 1999 the Chinese government has invested around 9% of their agricultural research budget in biotechnology.

Furthermore, they estimate that these investments represent around half of the developing world investments in agricultural biotechnology.

Biotechnology Research Capacity and Results

One measure of institutional capacity for biotechnology research can be examined by documenting their research outputs, such as the successful development of GM crop innovations. In this regard, a meeting was organized to review such progress, selecting 16 countries that reflect investment timeframes of over 10 years.

One of the initial findings of this meeting and study, referred to as *Next Harvest*, was that while intellectual property issues have thus far proven manageable for most countries, high regulatory costs found to ensure safety, increased public concerns, and lack of knowledge regarding the potential benefits and risks of biotechnology confine research products to the laboratory. This is seen when the number of research events in confined testing is compared with further requirements needed for advancement.

For these countries, (China, India, Indonesia, Malaysia, Pakistan, Thailand, The Philippines, Argentina, Brazil, Costa Rica, Bulgaria, Egypt, Kenya, Nigeria, South Africa, and Zimbabwe), significant investment has been made in biotechnology and regulation. Successful events, that are stable and transmit expression over generations, were clearly demonstrated from 76 public institutions. In this regard, 46 crops have been transformed, for a wide array of locally important traits. This is almost twice the number of crops transformed at CGIAR centers. In addition, germplasm being used and transformed is often local public lines, or other local breeding material, which increases the importance of such work.

Success of this national capacity can also be seen from the range of traits undertaken for transformation. From the

Next Harvest data, the most important traits include: insect resistance, virus resistance, herbicide tolerance, a combination of agronomic properties and product quality traits, and, fungal and bacterial resistance.

The actual genetic material incorporated also indicates wide access to and use of proprietary material, demonstrating research collaboration and capacity to insert and express the array of genes employed. However, final assessment of this work, and the opportunity to move towards outcomes, will be through products available to smallholder or local farming communities. This will of course be contingent on demonstration of safety, and appropriate mechanisms for the distribution of these new materials.

Among the six developing countries studied for research capacity and investments, each received government support for the establishment of biotechnology research centers, the creation of post-graduate training and research, and the formulation of regulatory frameworks. These efforts provide a good foundation for further development of biotechnology. However, a comprehensive strategy is lacking for biotechnology capacity development and expectations for this capacity as part of the larger agricultural knowledge system. With this in mind, the following findings and implications are presented.

1. *Successful implementation demonstrated*: Important progress has been achieved from the resources that have been available during the period studied. This includes progress in biotechnology research, and for biosafety and regulatory frameworks development. Existing capacity, while limited, has nonetheless completed important tasks. With increased management and better, more cohesive strategies, one would anticipate accelerated progress. This may mean that more funding is needed per researcher before building new capacity, especially if funding is not

available to sustain additional infrastructure and personnel.

2. *Building a strong public sector*: While capacity is evident, developing further management and analytical capacity is needed for public institutions that set agendas and priorities for biotechnology. These are particularly needed to ensure existing capacity serves rural community needs. Existing personnel will require further support as they are called to contribute to public awareness, and spearhead collaboration with the private sector on product development and diffusion.

   In addition to research capacity, the public sector or government is also responsible for regulatory decision making for biosafety. In fact, many research scientists go on to become regulatory officials, or part of the regulatory process, or serve on institutional biosafety committees. While much has been achieved in building regulatory capacity, further work is needed not only within the system, but to effectively deal with external pressures to the regulatory system from external factors, such as trade and markets. This additional capacity will strengthen the probability that efficient, timely and safe regulatory decisions can be made.

3. *Private sector relations*: In the countries reviewed, the creation of biosafety committees and the revision of property rights legislation should favour private sector participation in agricultural biotechnology research. The situation in China is different as the agricultural institutions in the public sector have promoted biotechnology innovation but commercialization has been hampered by conflicts with institutions in the public sector dealing with trade with regard to their intended objectives. There have been some initial public-private partnerships currently being implemented in China. This is generally not the case

in other developing countries. Participation by the private sector in the biotechnology innovation process remains limited. This is the case among all six countries during the period of the survey and for those surveyed as part of the *Next Harvest* study.

4. *Importance of priorities and research agendas*: Human capacity and impact from research will benefit from a strategy to foster biotechnology, including priority setting and research agendas. Governments need to make strategic decisions at the national level with a clear understanding of the costs, benefits and risks of biotechnology and its potential to help meet national goals. Policy makers need tools for planning and priority setting to help make informed trade-offs among commodities and research problems. There is also a need for more national and institutional commitment to fund agricultural biotechnology research and development. Strategic planning for future capacity building should be part of this process.

   These planning decisions are crucial to investments in biotechnology. There is no set rule that cuts across all countries and regions with regard to an appropriate level of funding and capacity for biotechnology. Each country determines its own level of investments, fitting its needs and capabilities. This should be part of a process that evaluates biotechnology research in the context of country priorities and needs, and also looks at alternative solutions.

   A decision to provide support requires each country to identify funding sources and evaluate their stability. As demonstrated above, most countries endured significant investment fluctuations and dependence on donors. If biotechnology is a desired innovation, then countries need to consider increasing domestic funding for high priority activities. For poorer countries, this may imply tight coordination between the public sector, donors, international research centers and the private sector. This coordination of roles does not have to be a formal process, but rather a response to an increasingly vibrant private sector in the country or region.

5. *Time for evaluation and review*: This report described GM crop outputs obtained from public research. While good progress can be seen from this work, it is time for more formal evaluations of capacity and investments. This is especially true as many of these programs have been running for over 10 years.

Biotechnology is a rapidly growing and changing endeavour, and capacity and investments must adjust accordingly. Also, international funding and research organizations have changes in their missions, programs and projects. The capability to analyse and adjust capacity for biotechnology research is essential to keep pace with current scientific advances and new research priorities. From this starting point, evaluation criteria were established for building biosafety capacity. Such methodologies can be considered for biotechnology research.

In addition, broader evaluations have been done for research and development originations globally. Lessons from this process, e.g. evaluating capacity development, are also crucial. This would require fine-tuning a well-established process for agricultural research to include or specifically tackle biotechnology. The goal here is that such evaluations contribute towards current capacity development planning, and to the organisation's performance.

REFERENCES

Byerlee, D. and K. Fischer., "Accessing Modern Science: Policy and institutional options for agricultural biotechnology in developing countries," *IP Strategy Today* No. 1, 2001.

Cohen, J.I. and R. Paarlberg., *Unlocking Crop Biotechnology in Developing Countries* – A Report from the Field World Development, in press, 2004.

Falck-Zepeda, J. B., G. Traxler and R. G. Nelson., "Cotton GMO adoption and Private Profitability," In: *Genetically Modified Organisms in Agriculture- Economics and Politics*, G. Nelson (ed.), London: Academic Press, 2001.

Falconi, C., "Agricultural biotechnology research indicators and managerial considerations in four developing countries," In: *Managing Agricultural Biotechnology – Addressing Research Program Needs and Policy Implications*, J.I. Cohen (ed.), Wallingford, U.K.: CABI Publishing, 1999.

Heston, A. and R. Summers., "The Penn World Table (Mark 5): An Expanded Set of International Comparisons, 1950-1988," *Quarterly Journal of Economics*, May 1991, pp.327-368.

Paarlberg, R., *The politics of precaution: Genetically modified crops in developing countries*, Baltimore, MD: Johns Hopkins University Press, 2001.

8

Research Priorities for Livestock Sector

Livestock in India is kept mainly by the small landholders and the landless that constitute bulk of the rural population. Thus, by being as an important means of income and employment for these households livestock helps alleviate poverty and smoothen income distribution. In addition, livestock asset can be easily converted into cash, and thus acts as a cushion against shocks of crop failure particularly in the less favoured environments.

Livestock makes substantial contributions towards conservation of environment. By utilising huge amounts of crop residues and byproducts as feed and fodder it contributes positively to the environment. In addition, utilisation of crop residues and byproducts by the animals makes substantial land available for food production, which otherwise would have been needed for fodder production.

Direct contributions of livestock in terms of supplies of draught power and dung save non-renewable environmental polluting energy sources. Thus, in view of such interactions between crop and livestock it is increasingly realised that livestock production would be more sustainable and growth-promoting in the mixed cropping systems as in India. Emerging economic forces are creating both challenges and opportunities for the livestock sector.

Perpetual growth in human population and sustained growth in per capita incomes are driving rapid growth in demand for animal food. There are estimates that by 2020 demand for milk, meat and eggs under different income and population growth scenarios would be in the range of 126-183, 6.3-12.1 and 9.5-18.5 million tons respectively. In the recent past, livestock outputs have grown at an annual rate of about 5 percent. If these trends were to continue future demand for various livestock products could be met from domestic supplies.

Serious doubts have however been expressed regarding sustainability of these trends because these have by and large been population-driven. Productivity of Indian livestock is low. For instance, cattle milk yield is about half of the world average of 2072 kg/animal/annum, and just 12-15 percent of the yield in the USA, Canada and Israel. Productivity of species other than dairy and poultry has been stagnating at an extremely low level. These figures suggest considerable scope for improving livestock productivity in India.

There is realisation that WTO regime will exert more pressure on livestock production to be efficient to meet the challenges of global competition. At the current level of productivity, export competitiveness of Indian livestock products is ambiguous. There is an anxiety that cheap imports of livestock products would threaten the Indian livestock industry. Further, there is increasing concern about the interaction between economic competitiveness and a number of social factors such as poverty, unemployment, food and nutritional security, and sustainability.

Livestock is a means of livelihood for millions of landless, marginal and small landholders who are often poor and this group would be the most vulnerable to the forces of economic competition. To meet the challenges of globalisation the key lies in continuous generation and dissemination of technologies that improve production efficiency of livestock. Well-targeted livestock research and development programs can meet the multiple challenges. Evidences indicate very high payoffs to investment in livestock research, which imply that there is enormous scope to invest in livestock research.

Unfortunately, the resources for agricultural research are limited and are becoming scarce. India spends only about 0.5 percent of AgGDP on agricultural research. This is below the average of the developed countries (2.5 %) as well as the developing countries (0.7 %). There is little optimism that, despite high rates of returns, investment in agriculture and livestock research will increase substantially at least in the immediate future. Nonetheless, there is scope to improve the growth of livestock sector by augmenting the productivity of different livestock species through better targeting of R&D efforts. That task necessitates objective articulation of priorities for research.

In recent years, priority setting has established itself an important tool in the management of limited research resources. It aims to improve strategic planning and accomplish efficient allocation of scarce research resources to improve research efficiency within the framework of national policy goals. Priority setting helps:

(i) review existing resource allocations,

(ii) update research agenda,

(iii) make resource allocations more transparent and unambiguous, and

(iv) strengthen credibility of the institution in soliciting support for research.

Historically, the process of agricultural research resource allocation in India has largely been based on informed scientific opinion concerning problems and opportunities. This involves a lot of subjectivity. The new problems and challenges are calling for more objective and transparent mechanisms to allocate research resources. The new challenges include expanding research system, chronic scarcity of resources for research and increasing severity and complexity of research needs. These necessitate a systematic assessment of priorities for livestock research.

In a seminal work, Jha, et al. generated quantitative estimates for allocation of agricultural research resources across regions and commodities. The study also provided useful insights for setting priorities for the livestock sector. The thrust of the study was on livestock products. In practice however species is the focus of research. This study is an attempt to build on the earlier attempts to assess research priorities for livestock sector for regions, species and commodities.

GROWTH OF LIVESTOCK SECTOR

Predominance of mixed crop livestock systems is one of the most important characteristics of Indian agrarian economy, wherein crops and livestock contribute towards the growth and sustainability of each other through input-output linkages. For the last few decades, both the crop and livestock sectors have been growing satisfactorily, the latter however has grown faster. Since 1970, livestock sector output has been growing at an annual rate of about 3.6 percent compared to pre-1970 growth rate of 1.3 percent.

The higher growth in later period was on account of market forces and technological change, and facilitated by increased availability of feeds and fodder, investment in animal health, and development of marketing network for livestock products.

Contribution of Livestock Sector

India possesses one of the largest livestock wealth in the world. In 1992, it had 205 million cattle, 84 million buffaloes, 115 million goats, 51 million sheep, 13 million pigs, 2 million equines and 306 million poultry birds (Table

1). And, the population of almost all the species has been growing.

Table 1. Structure and growth of livestock population in India

Species	*Population (millions) 1992*	*Annual compound growth (%) 1982-1992*
Cattle		
Male	101.6	0.40
Female	103.0	1.26
Total	204.6	0.82
Buffalo		
Male	17.3	0.66
Female	66.9	2.25
Total	84.2	1.90
Goat	115.3	1.93
Sheep	50.8	0.54
Pig	12.8	3.43
Horse	0.82	-0.96
Mule	0.20	5.88
Donkey	0.97	0.04
Poultry	306.2	4.52

The observed pattern of growth however indicates a shift towards economically more efficient species such as buffalo, pig, goat and poultry. Further, in case of bovines the incremental growth is less in population of males compared to females mainly due to substitution of animal draught power by mechanical power. This shows that livestock sector of India is both expanding and adapting to emerging socioeconomic, environmental and technological forces.

Contribution to AgGDP

Table 2 presents trends in contribution of livestock sector to gross domestic product (GDP) and agricultural gross domestic product (AgGDP). Agricultural sector contributed about 40 percent to the GDP in 1960s. This gradually decreased to 36 percent in 1970s, 31 percent in 1980s and 26 percent in 1990s.

Table 2.Contribution and growth of livestock sector, 1980-81 prices

	1971-80	*1981-90*	*1991-97*
Per capita GDP (Rs/annum)	1724	2177	2874
Share of AgGDP in GDP (%)	35.97	31.17	26.16
Share of livestock GDP in AgGDP (%)	na	17.60	21.38
Annual growth in GDP (%)	3.66	5.61	6.83
Annual growth in AgGDP (%)	2.43	3.29	3.99
Annual growth in livestock GDP (%)	-	7.31	4.92

The contribution of livestock to AgGDP however increased from 18 percent in 1980s to 22 percent in 1990s. Since 1980, livestock GDP has been growing at an annual rate of about 6 percent, which is higher than the growth in AgGDP and GDP. This indicates that livestock sector is likely to emerge as an engine of growth of the agricultural sector.

Inputs to Crop Sector

The contribution of livestock sector presented in Table 2 is underestimated because it does not reckon intermediate contributions of livestock. For example, draught power, which is an important intermediate contribution, is not reckoned while estimating livestock GDP.

Problems of Unemployment and Poverty

Livestock sector has considerable potential to contribute towards alleviation of problems of unemployment and poverty. About three-fourth of country's population lives in rural areas, and more than one-fourth of it is below poverty line. Poverty is largely prevalent among the landless, marginal and small farm households. In 1991-92, these households comprised 84 percent of the total rural households and shared one-third of the cultivated land. Average size of land holding is small, and has been declining continuously.

On the contrary, distribution of livestock is more egalitarian. Bulk of the livestock population is owned by marginal and small landholders; 71 percent of cattle, 63 percent of buffaloes, 66 percent of small ruminants, 70 percent of pigs and 74 percent of poultry. This implies that marginal and small landholders derive a considerable proportion of their income from livestock. Evidences indicate that increase in income from livestock in rural areas reduces income inequality.

Though, the contribution of livestock to AgGDP has been rising continuously, its contribution to rural employment is not so encouraging. In terms of principal activity status livestock employs about 5 percent of the rural work force. Its share however has declined to 3 percent in 1990s. Low share in rural employment is because livestock rearing in India is taken up as a subsidiary to crop production.

Food and nutritional security

Diet of an average Indian is cereal based. Lack of diversification towards nutrient-rich foods is considered to be the main cause of malnutrition and under-nutrition. In 1993-94, 30 percent of the population was undernourished. The problem is more severe to the population having less access to land for crop and livestock production. Animal food is rich in protein and other vital nutrients, and has potential to contribute towards alleviation of nutrition related health impairments. Consumption of animal products is low.

In 1993-94, per capita annual milk consumption was 51 kg, and meat 1.7 kg. Consumption of eggs was less than one egg/capita/month. These figures are much below than in many developing and developed countries. The demand for livestock products however is increasing as is indicated by the rising share of animal products in the food budget. In 1972-73, livestock products accounted for 14 percent of the food expenditure, and this gradually increased to 20 percent in 1993-94. This is expected to increase at a faster rate with sustained economic growth and attendant increases in per capita income.

Trade of livestock products

India has marginal presence in the world trade of livestock products. During triennium ending 1998, average export value of livestock products was Rs 13500 million. This comprised only 1 percent of the total export earnings and 6.2 percent of the agricultural export earnings. Meat and meat products are the main livestock products of exports accounting for over 90 percent to the total export earnings from the livestock sector. The export performance of livestock sector has improved during 1990s.

During TE 1998, India imported livestock products worth Rs 1877 million, comprising 0.1 percent of total imports and 1.5 percent of agricultural imports. Hides and skins comprised main items of imports. The dependence on imports however has declined. For instance, almost the entire milk demand now is met from domestic supplies. The trend in imports of hides and skins also has weakened.

Growth in livestock sector is important on many counts. First, by augmenting income and employment opportunities it would benefit millions of small landholders and the landless labourers who possess a sizeable proportion of livestock wealth. Thus, the growth in livestock sector is considered to alleviate poverty and lessen interpersonal income disparities. Second, increase in outputs of livestock would lead to increased consumption of livestock products, contributing towards lessening of nutrition related problems. Third, the past experience has shown that the increase in livestock production has facilitated import substitution of many livestock products.

Acceleration in growth would further reduce import dependence and facilitate export. Above all, being an integral part of agricultural economy, livestock would improve sustainability of the crop sector through provision of organic manure and draught power as inputs. India has huge livestock population, but with poor productivity. Cattle milk yield is half of the world average. The same holds good for beef and pork. Goat and sheep meat yields are less by 20-25 percent. The milk as well meat productivity of Indian livestock is less compared to average of the Asian countries.

The poor productivity is on account of scarcity of feeds and fodder, and poor animal health. Recent estimates indicate deficiency in dry fodder by 31 percent, green fodder by 23 percent, and concentrates by 47 percent. Further, a number of deadly diseases continue to occur frequently. Efforts to replace nondescript low yielding indigenous stock with high yielding crossbred stock too did not yield desired results. Only about 7.5 percent of cattle, 5 percent sheep, 14.5 percent pig and 32.9 percent poultry belonged to crossbred category in 1992. These figures indicate that technology would be a key factor in sustaining the growth of livestock sector in the decades to come.

History of Livestock Research

Livestock research in India is more than a century old. During the latter half of the nineteenth century, cattle plague was all pervasive and caused considerable loss to the livestock economy. The Government of India appointed the Indian Cattle Plague Commission (ICPC) in 1868 to report on the prevalence of rinderpest and other animal diseases, and to suggest measures for their prevention and control. The recommendation of this and another

Commission functioning during 1882-83 culminated in establishment of Imperial Bacteriological Laboratory- the fore-runner of the Indian Veterinary Research Institute in 1889 at Pune, Maharashtra.

In 1893, the Laboratory was shifted to Mukteshwar in Kumaon Hills, Uttranchal, and subsequently in 1913 to its current location at Izatnagar, Uttar Pradesh. Another landmark decision on the recommendation of the ICPC was the establishment of a veterinary college in Mumbai in 1883, and subsequently at Lahore, Kolkatta, Chennai and Patna. For over the next four decades or so, research efforts at Imperial Bacteriological Laboratory remained concentrated to develop preventive and curative methods to overcome the rinderpest menace.

In 1927, these efforts resulted into development of a highly efficacious vaccine for immunisation of cattle and buffalo against rinderpest. The research portfolio of the Laboratory diversified gradually to include other animal diseases. Among other successes, a major breakthrough was development of a vaccine against Ranikhet disease of poultry in 1940. The next important milestone towards strengthening livestock research was the establishment of Imperial Institute for Animal Husbandry and Dairying -the fore-runner of the Indian Dairy Research Institute at Banglore, Karnataka in 1923.

The mandate of the Institute was to impart training in animal husbandry and dairying for development of the dairy sector. In 1936, the Institute was renamed as Imperial Dairy Institute. After Independence, the Institute was relocated at Karnal, Haryana in 1955. The establishment of the Indian Council of Agricultural Research in 1929 on the recommendations of the Royal Commission on Agriculture was the most important landmark in the history of Indian agricultural research. The Council was mandated to promote, guide and coordinate agricultural and veterinary research in the country.

The mandate precluded the Council to maintain its own institutions, but to determine the research needs at regional and all India level and to support ad hoc projects at research institutes and state departments. The Royal Commission on Agriculture (RCA) had also made an in-depth analysis of the problems plaguing the growth and development of animal husbandry. The cattle were the main source of draught power for agriculture and milk for human consumption, but there was considerable imbalance between the two with a tilt in favour of the former.

With due consideration to this duality, the RCA recognised the need to improve milk yield of bovines and recommended improving milk production traits of cattle as well as buffalo through genetic enhancement of indigenous breeds by selection. The RCA also emphasized on the importance of improved feeding and management practices. With these recommendations, the focus of research of the Imperial Dairy Institute shifted towards improving milk yield. It may be noted that the RCA did not favour crossbreeding research because of fear of non-acclimatisation of crossbreeds to tropical climate of the country.

By late 1930s, artificial insemination (AI) technology has arrived in India. The isolated research efforts to try AI in cattle were started in 1939, but comprehensive studies on AI were initiated in 1942 at IVRI. It was found an effective tool to bring about genetic improvement in cattle and was recommended for extensive field application. At the same time, the livestock research portfolio diversified to include other animal species, that is sheep and poultry. The focus was on improving their genetic potential through crossbreeding.

Efforts to improve indigenous sheep through crossbreeding with exotic fine wool breeds had earlier been made in the beginning of the nineteenth century. These however could not sustain in the absence of proper organisations. Systematic efforts started in the late 1930s when ICAR initiated a number of pilot experiments for breeding superior sheep. Findings of these experiments evolved into a sheep breeding policy that emphasized improvement in indigenous breeds by selection in plains, and through crossbreeding in temperate Himalayan region and sheep tracts of Deccan plateau.

The beginning of poultry research in India can be credited to few Christian missionary organisations. Towards the end of nineteenth century, these organisations imported superior exotic breeds to improve the indigenous breeds through crossing. The organised efforts started in 1938 with the establishment of a Poultry Research Division in the Indian Veterinary Research Institute for undertaking research on poultry breeding, nutrition and health. The animal science research in the pre-Independence era was allocated fairly a good amount of funds.

In 1930s and 1940s, share of animal science research was about 30 percent of the total research funds of the ICAR. This laid a strong foundation for animal science research in the country. A review of research achievements by the ICAR towards the end of 1940s also revealed availability of a number of livestock technologies related to breeding, nutrition, dairy enterprise, sheep and wool, and poultry for field application. The research on many important species such as buffalo, goat, camel and pig was still lacking adequate attention.

Growth in Infrastructure

No major initiatives were taken to strengthen agricultural research in the early post-Independence years. However, a number of committees were constituted to suggest ways and means to strengthen the agricultural research system. On their recommendation, the first agricultural university was established at Pantnagar in 1960 with the help of ICAR and support and grants from United States Agency for

International Development. This was followed quickly in other states. At present, there are 28 agricultural universities, and most of these have faculty of animal sciences. There are two universities specifically meant for animal science research and education.

In 1950s, establishment of All India Coordinated Research Project (AICRP) on maize by ICAR was a major step towards strengthening agricultural research. During 1960s, 17 AICRPs were established but emphasis was on crop husbandry. However, during this period establishment of Central Sheep and Wool Research Institute at Avikanagar (Rajasthan) was a major step towards strengthening species specific research. Feed and fodder scarcity had repeatedly been pointed out to be the main limiting factor to improving animal productivity in the pre-Independence period.

The problem persisted in the post-Independence years also. This led to the establishment of Indian Grassland and Fodder Research Institute in 1962 at Jhansi, Uttar Pradesh with the objective of identifying fodder problems in different regions and research and management solutions thereof. Until 1966, agricultural research institutes were under the administrative and financial control of central Ministry of Food and Agriculture. ICAR was reorganised in 1966 and its mandate was revised to include 'undertake research' in the field of agriculture and animal husbandry. Thus, the research institutes under the Ministry of Food and Agriculture were brought under administrative, technical and financial control of ICAR.

Animal science research scenario changed drastically after mid-1960s. Milk production in the country had been stagnating around 20 million tons during the first two decades of post-Independence. By the end of 1960s, the Green Revolution technologies had resulted into tremendous success. Encouraged with this, a similar strategy was planned to increase milk production with emphasis on crossbreeding of indigenous cattle with exotic breeds and with simultaneous development of marketing facilities in the cooperative sector. The strategy culminated into initiation of the program 'Operation Flood' in early 1970s.

Animal science research infrastructure expanded considerably in 1970s and 1980s (Table 3). In early 1970s, a number of AICRPs were established. In late 1970s and 1980s, many species specific research institutes/centres were established by ICAR. At present, there are separate research institutes/centres for cattle, buffalo, sheep, goat, poultry, mithun, yak, camel and equine. Besides, there are institutes to conduct research on animal genetic resources, physiology and nutrition. In all, there are 2 national institutes, 5 central institutes, 5 national research centres, 1 national bureau, 2 project directorates and 17 AICRPs dealing with animal science research in the country. The larger institutes have also their regional centres and sub-stations in different agroclimatic zones.

Table 3. Number of animal science research institutes/centres ICAR.

	Upto 1970	*Upto 1980*	*Upto 1990*	*Upto 1998*
National Institutes	2	2	2	2
Central Institutes	1	3	4	5
National Research Centres	0	0	4	5
National Bureaus	0	0	1	1
Project Directorates	0	0	2	2
AICRPs	18	8	10	17

Several other public sector agencies such as general universities, Council of Scientific and Industrial Research, Department of Biotechnology, Department of Science and Technology, Indian Institute of Immunology, etc. and private institutions like Bhartiya Agro-Industries

Foundation also participate in livestock research directly or indirectly.

Financial Resources

The institutions under ICAR and the AICRPs supported by it conduct bulk of the basic and applied research on livestock. ICAR also asserts considerable influence in deciding the research portfolio of state research systems through its funding and advisory roles. Therefore, an overview of ICAR's resource allocation pattern is of considerable interest. The story however remains incomplete in absence of a profile of allocation patterns of state research systems. Nonetheless, the pattern of resource allocation in ICAR acts as a trendsetter for state agricultural research systems.

ICAR's outlay over the last three decades has increased considerably. In nominal terms, it increased from Rs 914 million in the IV FYP to Rs 33770 million in the IX FYP. However in real terms, ICAR outlay stagnated at around Rs 2000 million during the IV and V FYPs. It increased to Rs 3399 millions in the VI FYP and declined marginally in the VII FYP. It witnessed a quantum jump in the subsequent plans.

In the on-going IX plan ICAR outlay is 7651 million rupees. As percent of total outlay for agricultural sector ICAR's share has increased to 9 percent in the IX FYP for about 4 percent in the IV FYP. Research claims about 70 percent of the ICAR resources. Overtime, significant changes have occurred in pattern of ICAR resource allocation towards research. Its share in total ICAR outlay was around 60 percent in the IV and V FYPs. This increased to above 70 percent in the subsequent FYPs.

In relative terms, ICAR gives adequate emphasis on livestock research. In the IV and V FYPs, ICAR allocated about 27 percent of its total research outlay to livestock research, which was about 1.5 times higher than the share of livestock in total value of agricultural output. It however fell drastically to 14 percent in the subsequent two FYPs. In the VII FYP, it increased to 18 percent and further to 19 percent in the IX FYP.

IMPACT OF LIVESTOCK RESEARCH

Livestock research has generated a number of technologies in the areas of animal breeding, nutrition, health and management. Adoption of these technologies has remained limited and sporadic. The value of output of livestock sector has however increased substantially since 1970. This could be attributed to the technological change as well as to improvements in efficiency of livestock production.

The joint impact of technological change and technical efficiency can be captured by the growth in Total Factor Productivity (TFP). TFP index suggests that prior to 1970, growth in livestock output was input-driven. However since 1970, TFP index grew faster and accounted for about 37 percent of the output growth. TFP measures the joint impacts; it does not indicate explicit contribution of either technological change or technical efficiency. Considering the level and rate of adoption of crossbreeding technology, contribution of the former to output growth appears to be limited.

Between 1982 and 1992, share of crossbred cattle in total cattle population increased from 4.7 percent to 7.4 percent, improved poultry from 21.7 percent to 32.9 percent, crossbred pig from 9.5 percent to 14.5 percent, and sheep 3.3 percent to 5.0 percent. Since early 1970s, milk output of cattle and buffalo has been growing at an annual rate of 5.2 and 4.4 percent respectively. The growth in milk yield was 3.2 percent in case of cattle, and 1.9 percent in case of buffalo.

Meat output too grew at a rate above 4 percent a year. The rate of growth however varied across species; highest growth occurred in beef and veal, followed by pork, poultry meat, buffalo meat, goat meat and mutton. The growth has occurred due to growth in number of animals slaughtered. The yield growth was negligible in case of most of the species. These trends indicate that technology has made limited contribution to the growth of livestock sector. Nonetheless, there exists considerable scope to raise productivity of livestock through focussed technological interventions.

METHODOLOGY

Livestock species are multi-functional in nature i.e. they generate multiple/joint outputs. Each of the output has economic value, though its relative utility may vary over space and time. Multi-functionality is more pronounced in the subsistence oriented mixed crop livestock production systems as in India, where livestock is maintained not only to produce food but also to provide draught power and organic manure to agriculture.

Mixed crop livestock production systems are dynamic, and possibilities of tradeoffs between crops and livestock as well as within the functions of livestock are quite high. For example, with the emergence of energy intensive cropping systems in India the emphasis of cattle breeding research is gradually shifting from dual-purpose breeds to milch breeds.

The multi-functionality thus makes priority assessment exercise a complex process. Besides, livestock research is long-term, capital-intensive, uncertain and slow in producing benefits, which bring in elements of uncertainty in priority setting. Over the last three decades, a number of priority setting methods have been developed. These differ widely in scope of analysis, degree of sophistication and applicability. The priority setting methods can be classified based on the objectives of priority setting, measurement concepts and the time dimension.

Broadly, there are six methods of priority setting that have been widely applied in empirical research:

(1) Scoring or weighted criteria model,

(2) Economic surplus/benefit-cost models,

(3) Mathematical programming models,

(4) Simulation models,

(5) Econometric methods, and

(6) Analytical hierarchy process.

Some investigators have also used a combination of scoring and economic surplus approaches in setting the research priorities. Each of these approaches has its advantages and limitations.

Choice of Method

The choice of method depends on the objectives of research, level of priority assessment, and its simplicity in application, data requirement and capacity to allow participation of stakeholders in priority assessment exercise. Multi-criteria scoring model has been used in this exercise. This is often known as congruence method. This incorporates multiple objectives for making choices. The objectives included here are growth, equity, sustainability and participation in trade. These are complex and often argumentative in nature.

Scoring model incorporates multiple objectives by modifying the simple and traditional measure of research evaluation—changes in value of production- to consider the concerns of equity, sustainability and trade. Given the relative importance of objectives, scoring model makes tradeoffs between objectives explicit. The approach is relatively transparent and allows extensive and active participation of stakeholders. Finally, the approach is simple to apply and is not data intensive.

The model can be used to rank a long list of commodities, research alternatives and target domains.

Main limitation of this approach is its simplicity and flexibility. Often, there is a tendency to add in more and more criteria without checking for their internal logic and consistency. This gives rise to duplicating criteria. This happens as most of the criteria are directly or indirectly related with the efficiency criterion- the main objective of research. Further, scoring model allocates research resources across regions or commodities in proportion to their contribution to the gross value of production.

This implicitly assumes equal opportunities for research across commodities and regions. That means the value of new research generated is proportional to the value of output from the sector. The analysis is based on present values and presumes that relative contributions are constant. Thus, the results provide only the starting point for research priority setting. It also does not account for research spillovers and domestic and trade policies. The results of scoring model however can be improved by combining them with other methods such as economic surplus and mathematical programming that consider probability of research success, adoption rate, research spillover effects and distribution of research benefits.

Steps in Scoring Model

Scoring model involves seven broad steps to arrive at a comprehensive assessment of the priorities. These are:

- Identification of research objectives and their indicators (extensity parameters)
- Assigning appropriate weights to research objectives
- Selection of research priority dimensions
- Construction of initial baseline (IBL)
- Modification of IBL
- Construction of final baseline (FBL)
- Assessment of research priorities

Identification of research objectives

The first step is to identify research objectives that are consistent with the national or regional development goals. The goals often include growth acceleration, improvement in equity, sustainability of natural resources/production systems and participation in international trade. The 'growth acceleration goal' suggests that research should enhance productivity in a cost-effective manner.

Equity implies distribution of efficiency gains across regions, socioeconomic groups, farming systems and so forth. Sustainability is concerned with conservation and/or judicious use of natural resources. Participation in trade indicates that research needs to focus on commodities having potential for export promotion or import substitution. These developmental goals have been the guiding principles for agricultural research system in defining the research objectives.

The prominent research objectives corresponding to the national goals outlined above are: enhancing agricultural productivity, augmenting income of people living below poverty line and improving their nutritional status, increasing carrying capacity of land and accelerating export growth. The next step is to identify the indicators/extensity parameters reflecting the magnitude of the problems to be addressed by the research system. Accordingly, the research activities should target the regions where the magnitude of the problem is large.

Value of livestock production (VOP)

Research enhances production potential of livestock, and benefits of the research thus can be increased manifold by large-scale adoption of its outputs. That means value of research is proportionate to the value of production, and the latter therefore reflects the research objective of increasing productivity or production efficiency.

In this analysis, livestock species has been considered as the focus of research and thus we have included the triennium ending 1997-98 average value of the outputs of each species. Value of the following outputs is available to estimate the value of contribution of a species: total milk (cattle+buffalo+goat), cattle meat, buffalo meat, goat meat, sheep meat, pig meat, poultry meat, eggs, cattle hide, buffalo hide, goat skin, sheep skin, sheep wool, goat hair, pig bristles, camel hair, total dung (cattle+buffalo).

The value of some byproducts such as bones, blood etc. is also available but not by species. However, these comprise only a small fraction of total VOP of livestock and hence have not been considered to compute the contribution of individual species. In several cases, we had to apportion the values by species. For instance, total value of milk has been apportioned by species (cattle, buffalo and goat) taking into consideration the milk output and price of milk of different species. Similarly the total value of dung has also been disaggregated at the species level considering their populations and dung evacuation rates. Further, the values of draught services of different species were not available, so we had to estimate these values.

The value of draught power contribution of livestock has been estimated using information on draught power utilisation from the Cost of Cultivation Studies. The Cost of Cultivation reports provide per ha value of animal draught power for major crops by states. A single estimate of value of animal draught power per ha was arrived at for major crops in different states.

Thus the total value of animal draught power was arrived at by multiplying the per ha value with gross cropped area in a state. This was further disaggregated at species level in proportion of horsepower available from different draught species. The species are cattle, buffalo, goat, sheep, pig, poultry, camel and equine. Other species such as mithun and yak could not be included in the analysis due to lack of information on their outputs and services. These species are important in certain agroecological regions, but are minor to the livestock sector.

Number of undernourished persons

The number of persons below poverty line and the number of undernourished persons are indicators of economic inequality and nutritional insecurity respectively. As distribution of livestock wealth is more equitably distributed compared to land, it is expected that strengthening livestock research in the regions having comparatively more number of the poor and undernourished would help reduce interpersonal and interregional economic disparities and reduce incidence of malnutrition and under-nutrition.

Common property resources

Common property resources contribute significantly to the feed and fodder consumption and thus to the sustainability of livestock production. The contribution is higher in case of landless, marginal and small landholders who own bulk of the livestock wealth. The common grazing resources in India have however been deteriorating in quantity and quality. Therefore, conservation and management of CPRs is necessary to protect the interests of these households and to sustain the growth of livestock sector.

The focus of livestock research thus should be relatively higher in the regions having more grazing and wastelands. In this analysis, we have considered area under barren and unculturable lands, permanent pastures and grazing lands, culturable wasteland, fallow lands and lands under miscellaneous tree crops to constitute the CPRs for the livestock sector. Triennium ending 1997-98 average of the CPR area has been used in the analysis.

Livestock exports

Livestock sector export constitutes around one percent of total export earnings. There is enormous scope to improve the contribution of livestock products in export. In the past, India has been a net importer of various livestock products particularly powdered milk. The dependence however has declined drastically in recent years on account of tremendous growth in milk production. Besides, India has the advantage of producing milk as well as other livestock products at lower costs because of availability of cheap labour.

Further, trade liberalisation is opening up opportunities for improving export performance of livestock products. Participation in trade therefore has been considered as one of the criteria for priority setting. The indicator for this is the mean value of livestock products exported during triennium ending 1997-98. These parameters quantify the magnitude of the problem to be addressed by the research system, and imply that if the research system has to be efficient in achieving specific objective(s)—enhancing productivity, alleviating poverty, improving nutritional security, improving sustainability or promoting exports, it should target regions having comparatively high value of the corresponding indicator.

Appropriate weights to research objectives

The next step is to assign weights to the indicators of research objectives. This is needed to construct a composite index for priority ranking of regions, species and commodities. Each indicator is assigned an appropriate weight depending on the relative importance of the concerned research objective. There is no a priori information available on weights to different national developmental objectives. Thus, equal weights (0.25) have been assigned to each of the objective.

In case of more than one indicator of an objective, each one has been given equal weight without altering the overall weighting structure. For example, the number of poor, and undernourished population, the two indicators of the equity, share half of the weight (0.125) assigned to the equity objective (0.25). The methodology however permits use of varying weights to reflect the relative importance of different objectives, and a sensitivity analysis has been done to assess the impact of differing weighting schemes.

Research priority dimensions

Livestock research prioritisation has multiple dimensions. Priority indices can be developed for regions, livestock species and commodities or a combination of these. In this exercise, we have considered regional, species and commodity dimensions. Regional dimensions ensure that benefits of research are equitably distributed across regions. Since R & D efforts are generally targeted to homogenous agroclimatic regions, priority assessment must consider regional dimensions.

In India, agroclimatic regions cut across administrative boundaries and the necessary information for priority assessment is not available for the livestock sector at this level. Therefore, the state has been taken as the unit for analysis. All 25 states and 7 union territories have been included in the analysis. The regional research priorities can be translated into species priorities. Species included in this exercise are cattle, buffalo, goat, sheep, poultry, pig, camel and equine. As the animal in the mixed farming systems performs multiple functions, the priority for a species is further decomposed into functional priorities.

Initial Baseline Modification

Value of IBL is a cumulative product of weights and shares

of extensity parameters of research objectives. Research resource allocation based on IBL may not fully capture the major concerns of growth, equity, sustainability and exports, as this reflects size of the problem, but ignores its intensity. For example, a state might be ranking very high in terms of its share in the absolute number of the poor in the country it may not have the same ranking in terms of intensity of poverty.

In research prioritisation, intensity of the problem is also important. The effect of intensity of the problem cannot be captured in an analysis based on extensity parameters alone because of the problem of aggregation of values of intensity parameters across states. The intensity values are thus used to modify the IBL to arrive at the final baseline (FBL), which eventually reflects both the size and intensity of the problems. The intensity indicators are also known as modifiers.

Modifiers

Eight intensity indicators were identified for modification of the initial baseline. These include scope for growth in milk production (yield gaps in indigenous cattle, crossbred cattle and buffalo), per capita state domestic product and combined share of landless, marginal and small farmers in total livestock population (adult cattle equivalent), per capita availability of milk and eggs, and livestock density (adult cattle units/ha of net cropped area plus land available for livestock).

Efficiency modifiers

There is considerable variation in milk yield across species and states. Yields of different breeds of cattle and buffalo obtained under experimental conditions indicate possibilities of increasing milk yield. Due to lack of prior information on potential milk yield of the dominant breeds in different states it is assumed that the local breeds in a state would be replaced by the high yielding breeds of the same species. For this purpose, yield of Red Sindhi for indigenous cattle, of Haryana X Fresian for crossbred cattle and of Nili Ravi for buffalo are taken as yardsticks.

The difference between the yield of these breeds obtained under experimental conditions and the mean of the realised yield of all breeds of the respective species provide estimates of yield gaps. This however is a crucial assumption as the performance of high yielding breeds varies across states depending on agroecological and socioeconomic conditions. Nonetheless, the gap between realised and potential yield exists even for the dominant breeds in different states.

While yield gaps are attributed to differences in input use and management practices, magnitudes of such gaps are important determinants of strategic or applied research. Thus, greater the gap between potential and realised yield higher will be the scope for production growth through applied research. In such regions, the technology is available but its adoption is constrained due to biotic, abiotic and socioeconomic factors. Therefore, in such situations there is a greater need for socioeconomic research focussing on identification of constraints to adoption of improved technologies. In states like Punjab and Haryana where the realised yields are nearer to the potential yields, emphasis has to be on strategic research to break the biological yield barriers.

Equity modifiers

Four modifiers viz. per capita state domestic product, share of landless, marginal and small farmers in livestock population, per capita availability of milk and per capita availability of eggs have been used to address equity concerns. The first two address the economic equity concerns, while the latter two address the issues related to nutritional equity. To improve income distribution more research resources are needed for the states with lower per capita income and higher concentration of livestock among landless, marginal and small landholders. This would help improve upon the interpersonal and interregional disparities.

Incidence of undernourishment is expected to be lower in states with higher per capita availability of livestock products. Diversification of food basket towards livestock products would help mitigate problem of undernourishment. Therefore, livestock research should target regions having higher incidence of undernourished population. Milk and eggs are the preferred livestock products in the daily diet of the majority population. Thus, per capita availability of these commodities as modifiers takes care of the intensity of nutritional insecurity.

Sustainability modifiers

With an annual growth rate of about 5 percent livestock sector is one of the fastest growing sub-sectors of Indian agriculture. It is however unclear whether this trend will sustain considering the growing number of livestock, declining per capita land availability, slow growth in area under fodder crops and quantitative and qualitative deterioration of common grazing lands.

In other words, land and other natural resources available for livestock would come under heavy pressure, threatening the sustainability of natural resources as well as livestock output growth. This calls for emphasis on yield-enhancing research that would help optimise livestock population commensurate with available resources. Livestock density during the triennium ending 1997-98 is used as a modifier to address the sustainability issues in livestock production.

Weights and Signs of Modifiers

Having identified modifiers, the next step is to assign

weight to modifiers and determine their signs as to quantify their individual or joint impact on IBL. The magnitude of the weight of the modifier directly influences the relative emphasis of each concern, and this is the explicit weight of the modifier. Besides, the variability in distribution of a modifier across states exerts indirect influence, which is termed as the implicit weight of the modifier.

Recognising the role of implicit weight of the modifier in impacting the IBL we have not introduced varied explicit weights. Evidences indicate that higher weights to the modifiers bring in considerable distortions in relative ranking of regional priorities. Thus, equal weights are assigned to all the modifiers. The methodology however is flexible to accommodate varying weights. A modifier will have either a positive or negative impact on initial baseline. For instance, higher the per capita domestic product in a state, lower should be the research emphasis there to reduce interregional disparities in income distribution. Therefore, a negative sign is attached to this modifier.

On the other hand, inter-household distribution of livestock would carry a positive sign as this helps improve interpersonal equity in distribution of benefits of livestock research. Modifiers of nutritional equity would have negative signs as the states having less per caput availability would need more research resources to enhance the production potential of livestock and thereby the availability of different livestock products.

Positive signs are attached to efficiency modifiers as the states with higher yield gaps need more emphasis to enhance the yield level nearer to the potential yield. The livestock density variable, which addresses sustainability concerns, carries a positive sign. Thus, signs of the modifiers should be appropriately decided to quantify their impact on IBL, as the selection of signs involves logical judgements. The methodology enables use of different signs to different modifiers to evaluate their impacts and tradeoffs involved in priority setting with multiple objectives.

PRIORITIES IN LIVESTOCK RESEARCH

Livestock sector accounts for nearly one-fourth of the gross value of agricultural output in the country. This however varies widely across states. In Punjab, Jammu & Kashmir and Meghalaya, livestock contributes more than one-third to the total value of output of agricultural sector. The states in which the contribution of livestock ranges between one-fourth to one-third to the gross value of agricultural output are Bihar, Haryana, Himachal Pradesh, Rajasthan and Nagaland.

In other states, except Goa, Orissa, Arunachal Pradesh, Assam and Tripura, it ranges between 20 to 25 percent. Estimates by Jha et al. suggested an allocation of 23 percent of the agricultural research resources for animal science research. Across states, this varied from 16 percent in Assam to 34 percent in Rajasthan, which was largely in conformity with the contribution of livestock to gross value of agricultural output in these states.

Extensity Parameters

Distribution of extensity parameters viz. value of production (VOP), poverty (POOR), undernourished population (UNUR), sustainability (CPR) and exports (EXPO) across states.

Value of output of livestock (VOP)

VOP is the initial indicator of research resource allocation across regions when the sole objective of research is to improve the efficiency of research. This means distribution of research resources should be in proportion of their shares in total value of output produced in the country.

Poverty

The distribution of poor people across states shows their highest concentration (18.9%) in Uttar Pradesh. Bihar ranks second with a share of 15.2 percent, and is followed by Maharashtra (10.3%), Madhya Pradesh (8.2%), West Bengal (8.1%) and Tamilnadu (6.6%). This indicates that emphasis on poverty alleviation would favour more allocation of resources to these states.

Undernourished population (UNUR)

Uttar Pradesh has highest undernourished population (12.2%) in the country. This is closely followed by Maharashtra (12.1%). Tamilnadu, Bihar, Andhra Pradesh and Madhya Pradesh are other states having sizeable share in undernourished population. Concerns for nutritional security/equity would favour higher resources for these states.

Common property resources (CPR)

The regional distribution of CPRs highest concentration in Rajasthan (19.6%). Andhra Pradesh, Gujarat, Maharashtra, Uttar Pradesh and Bihar too account for substantial share in total CPRs. Thus, these states should receive higher priority if the emphasis of research were to improve sustainability of livestock systems through conservation and management of CPRs.

Exports (EXPO)

This indicator is designed to address the trade-related issues in priority assessment. It is defined as the share of a state in national export earnings from the livestock sector. The export portfolio includes live animals, dairy products, meat, and poultry products. With export as a criterion of allocation of research resources, bulk of the resources are shared by Maharashtra, Uttar Pradesh, Andhra Pradesh and Bihar in that order.

Initial Baseline (IBL)

The distribution of extensity parameters indicate that research emphasis should be on the regions having comparatively higher share in value of output, poor and undernourished population, grazing resources and exports. The shares of the states however vary for each extensity parameter. A composite index of allocation of research resources has been generated by assigning equal weight to each specified objective. In case of equity objective, the weight is shared equally between its two indicators, i.e. population of the poor and the undernourished.

VOP based allocation assigns high priority to Uttar Pradesh, Maharashtra, Punjab, Madhya Pradesh, West Bengal, Andhra Pradesh, Rajasthan and Bihar. When poverty is superimposed on VOP index, Uttar Pradesh consolidates its position and Bihar gains considerably by moving to second place in the priority ranking. Maharashtra, Madhya Pradesh and West Bengal gain in priority order, while Punjab, Andhra Pradesh and Rajasthan lose their positions.

With addition of nutritional security dimension, Uttar Pradesh continues to rank first but loses marginally. Maharashtra reverts to its original ranking to second place and Bihar slides to the third place. Madhya Pradesh, West Bengal and Andhra Pradesh are other states in the priority order. Punjab loses considerably and is replaced by Tamilnadu. The concerns for equity thus bring in considerable trade off in regional research resource allocation. The priority ranking of the states is further reshuffled on incorporation of sustainability considerations. Uttar Pradesh maintains its position, but with considerable decline in its share in research resources.

Rajasthan emerges second in the priority list, and is followed by Maharashtra, Bihar, Madhya Pradesh, Andhra Pradesh, Gujarat and Tamilnadu. The final iteration with all the extensity parameters (including exports) suggests that nearly three-fourths of the livestock research resources would be claimed by Uttar Pradesh (13.9%), Maharashtra (12.5%), Andhra Pradesh (9.7%), Bihar (9.6%), Rajasthan (8.1%), Madhya Pradesh (6.9%), Tamilnadu (5.3%) and Gujarat (5.2%). The IBL/VOP ratio indicates that in the IBL based allocation all the states in eastern (except West Bengal) and northeastern regions improve their shares over VOP based allocation. Among the western states, Maharashtra and Rajasthan gain considerably, and Madhya Pradesh and Gujarat lose marginally.

In the southern region, Andhra Pradesh and Tamilnadu improve their shares over the VOP based allocation, and the shares of Kerala and Karnataka get reduced at the margin. In the northern regions, Punjab, Haryana, Jammu & Kashmir lose considerably, while Uttar Pradesh loses marginally over their VOP shares. It is only Himachal Pradesh in this region that gains marginally over its VOP based share. The considerations of poverty, nutritional security, sustainability and exports help these states to gain over simple VOP based allocation.

Final Baseline (FBL)

The IBL has been modified to accommodate the intensity of the problem to be addressed by the research system. The impact of the modifiers on the IBL, and modified base line. The distribution pattern of livestock research resources changes when extensity indicators and their modifiers are considered together. Shares of Haryana, Punjab and Gujarat, Andhra Pradesh, Karnataka, Tamilnadu, Goa, Maharashtra and Rajasthan further decline, while other states consolidate their positions over IBL based allocations.

This implies that neglect of intensity dimensions may lead to sub-optimal allocation of research resources.

These tradeoffs become more pronounced when FBL based allocation scheme is compared with VOP based allocation scheme. The FBL/VOP ratios reflect these. In the absence of tradeoffs among research objectives, FBL/VOP ratio would be closer to unity. A ratio greater than unity for a state implies that concerns for equity, sustainability and exports favour more allocation of resources for the state. Vice versa, a ratio less than unity implies less research emphasis for the state.

The FBL/VOP ratio lies between 0.95 and 1.05 for Karnataka and Uttar Pradesh. This implies that their shares in total livestock research resources remain almost same whether resources are allocated based on economic efficiency criterion alone or in combination with equity, sustainability and export criteria.

Table 4. Tradeoff in regional research priorities

Ratio (FBL/VOP)	*States*
> 1.50	Goa, Manipur, Meghalaya, Mizoram, Nagaland, Sikkim, Orissa
1.25-1.50	Andhra Pradesh, Arunachal Pradesh, Assam, Bihar, Maharashtra, Tirpura
1.05-1.25	Himachal Pradesh, Tamilnadu, Rajasthan
0.95-1.05	Karnataka, Uttar Pradesh
0.75-0.95	Gujarat, Kerala, Madhya Pradesh
< 0.75	Haryana, Jammu & Kashmir, Punjab, West Bengal

The FBL/VOP ratios for other states indicate that concerns for equity, sustainability and export lead to considerable regional tradeoffs in resource allocation. With these concerns, northeastern and eastern states gain over the VOP based allocation. In the southern region, Andhra Pradesh and Tamilnadu consolidate their positions in the FBL based scheme, while Kerala loses. In the western region, Goa, Maharashtra and Rajasthan would receive higher priority in allocation of incremental research resources, while Gujarat and Madhya Pradesh would lose.

In the north, Himachal Pradesh would be benefited in the FBL scheme, while Punjab, Haryana and Jammu & Kashmir would lose. The tradeoffs are quite large for some states. The FBL based readjustments in resource allocation indicate considerable reduction in shares of Punjab (72%), Haryana (68%), Jammu & Kashmir (41%), West Bengal (35%) and Kerala (20%); and a moderate reduction in shares of Gujarat (12%) and Madhya Pradesh (9%). The shares of northeastern states however increase considerably.

The increase is 28 percent for Assam, 40 percent for Arunachal Pradesh and Tripura, 122 percent for Meghalaya, 174 percent for Sikkim, 188 percent for Manipur, 210 percent for Nagaland and 274 percent for Mizoram. Orissa, Goa, Bihar, Maharashtra and Andhra Pradesh too gain 30 percent or more in the FBL based allocation.

It may be noted that most of the states that gain in FBL allocation are backward or are in the development stage of agriculture, while those lose or maintain their shares in incremental allocations are in the fairly advanced stage of development. Further, in the former group of states contribution of livestock is less compared to the latter.

The results imply that though improving production efficiency is the main emphasis of research, ignoring long-term social and environmental goals in allocation of livestock research resources might aggravate regional disparities in livestock development. Unfortunately, no reliable information is available on the investments made in animal science research at the state level to verify the readjustments suggested by this analysis.

SPECIES PRIORITIES IN RESEARCH

The main objective of the animal science research is to accelerate the growth of livestock sector by improving the productive potential of the species. A species therefore is the unit of research, and allocation of research resources at this level has operational significance. Since an animal species has the characteristic of producing multiple/joint outputs, it is also important to bear in mind the relative utility of different products and services provided by a species while assessing research priorities. In this exercise, we have first assessed species priorities, and then identified priorities for different functions of the species.

Species Priorities in India

The FBL calculations accord highest priority to buffalo research with a share of 40.2 percent in the total research resources. Cattle with a share of 37.6 percent follow closely. Though, buffaloes are considerably less in number than cattle, higher priority to buffalo research is due to their higher milk yield and premium price on milk due to higher fat content. Poultry, goat, sheep, equine, pig and camel are other species in the priority order. Small ruminants and poultry demand 10 and 9.6 percent of the research resources respectively.

No significant tradeoffs in resource allocation are observed when all objectives are taken together. The FBL/VOP ratio reflects this. A ratio of greater than unity for a species indicates additional emphasis on this at the cost of the species for which FBL/VOP ratio is less than unity. Thus, FBL/VOP ratio suggests added emphasis on pig, goat and poultry (FBL/VOP ratio > 1.05). The FBL/VOP ratio for cattle, buffalo and sheep lies between 0.95 and 1.05, and implies little tradeoffs in resource allocation due to objectives other than efficiency.

With considerations of equity and sustainability, the share of poultry remains almost the same as in VOP based allocation. While consideration for export make poultry research gain a little. Camel and equine however lose their share in the FBL based scheme. Since dairy animals rank higher in the prioritisation, so are their products/services in commodity prioritisation. Milk production research claims 68.3 percent of the livestock research resources, followed by draught power (8.7%), poultry meat (7.2%), small ruminant meat (6.3%), eggs (3.6%) and beef and buffalo meat (2.4%).

The research on other products and byproducts (hides, skins, wool and hair) does not demand much of the resources. Having decided species priority, we determine priorities for products and services that a species provides. Meat and byproducts should receive the rest. In case of buffalo, most of the resources should be spent on milk production research. Draught power, meat and hides share the rest. Meat production emerges as the priority research area for small ruminant sector. The relative commodity shares however are different for sheep and goat. More than half of the goat research resources need to be allocated for meat and followed by milk. While in case of sheep, three-fourths of the research resources should be earmarked for meat research. Wool research claims only 11.4 percent.

Species Priorities in States

Economic contribution of different species varies across states. So are their shares in research resources. In northern states, buffalo research emerges as the main priority. In Haryana and Punjab, cattle and poultry are next to buffalo in priority order, while in Uttar Pradesh it is the cattle and goat. Cattle and buffalo are equal candidates for incremental research resources in Himachal Pradesh. Priorities in Jammu & Kashmir favour cattle, buffalo, sheep and poultry.

In southern region, species priorities vary considerably across states. For example, in Andhra Pradesh buffalo and poultry appear as the main priority species for livestock research. Cattle, sheep, goat, equine and pig are next in priority ranking. In Kerala, on the other hand, cattle research claims bulk of the livestock research resources, and is followed by poultry, buffalo and goat. Cattle and buffalo demand almost equal attention (41% each) in Karnataka, and poultry with a share of about 8 percent comes next in priority. In western region, cattle and buffalo research should receive almost equal priority, with the exception of Gujarat where the buffalo research needs to receive almost twice the resources allocated to cattle research.

Poultry research is important in Goa and Maharashtra. Issues related to goat too constitute an important research agenda in Rajasthan, Gujarat and Madhya Pradesh. In eastern region too, species priorities vary considerably across states. In Bihar, buffalo and cattle claim 34 and 31 percent of the total livestock research resources respectively. Goat research also demands considerable resources (21%), and is followed by poultry (8%).

Cattle research is important in Orissa (59%). Buffalo, poultry and small ruminants need equal emphasis (13-14 %) here. In West Bengal also, cattle receives highest priority, and goat and poultry follow in the order. In northeastern region, highest priority should be accorded to cattle research (53-76 %). Poultry is the next most important priority in some states (Assam, Arunachal Pradesh, Manipur and Tripura), while pig is in others. These results indicate that dairy animals (cattle and buffalo) would continue to attract significant attention in the livestock research cutting across regional boundaries though the relative emphasis varies across states.

Buffalo is the main priority species in most of the northern states; cattle in majority of the eastern and northeastern states; and both cattle and buffalo in the western states. In southern region, cattle should constitute the main research agenda. Further, livestock research agenda need to take care of the emerging diversification trends in the livestock sector. For instance, poultry and pigs in the northeastern regions, poultry in the western region and goat in the eastern region contribute substantially to the gross value of livestock production and should receive adequate attention in the process of research resource allocation.

Product Priorities by Species

After deciding the pattern of research resource allocation by species in different states, we assess the commodity priorities in the species research. The distribution of functional priorities for different species would vary across states due to differences in the functional utilities of a species. A regional comparison of the suggested allocation of cattle research resources shows emphasis on milk production research in most of the states. The emphasis however varies across states. Except in Andhra Pradesh, Goa, Gujarat, Madhya Pradesh, Orissa and most of the northeastern states milk production research claims more than three-fourths of resources allocated to cattle research.

In Andhra Pradesh, Gujarat, Madhya Pradesh, Rajasthan, Orissa, Arunachal Pradesh, Assam, Mizoram, Sikkim and Tripura, draught power research also needs considerable emphasis. Beef research is important in most of the northeastern states. For buffalo research, a pattern of commodity prioritisation similar to that at all India level is observed for most of the states.

At all India level, the analysis indicates research concentration on milk. In Kerala, Orissa, West Bengal and few northeastern states draught power and/or meat research is also important. Sheep research in most of the states should concentrate on meat production. Exceptions are Himachal Pradesh and Rajasthan where wool research emerges as the main priority. Poultry has been emerging as an important sub-sector of livestock in India. This is in response to growing demand for poultry products. At the national level, poultry research resources should be allocated in the ratio of 2:1 between meat and egg research. This however varies across states.

The priority in Haryana, Himachal Pradesh, Jammu and Kashmir, Kerala, Goa, Rajasthan, Bihar, Orissa, West Bengal and northeastern states by and large corresponds to the national priority. In Punjab, Andhra Pradesh, Gujarat and Maharashtra the ratio approximates to 3:1. Egg production research gets precedence over meat production research in Uttar Pradesh, Karnataka, Tamilnadu, Madhya Pradesh and Sikkim. Meat and bristles are two important outputs of pig, the former however should comprise the focus of pig research in all the states.

In Jammu & Kashmir, Tamilnadu, Gujarat, Madhya Pradesh, Andhra Pradesh, Orissa and Haryana research on bristles too demands attention. Besides males of cattle and buffalo, camel and equine are other important sources of draught power. Camel population is concentrated largely in Rajasthan, Gujarat, Haryana, Uttar Pradesh and Madhya Pradesh where it is used for both farm and non-farm activities. Camel also provides hairs, which are used in carpet industry. The main emphasis of camel research however should be on improving draughtability.

Equines are found mainly in Uttar Pradesh, Rajasthan Haryana, Himachal Pradesh and Punjab, and provide draught services mainly to the non-farm sector. The analysis indicates considerable regional variation in priorities for outputs and services of a species. In irrigated states, improving milk yield of cattle and buffalo is the main research agenda, while in rainfed states research on draught bovines also demands considerable attention. In case of other species too, there are marked regional differences in preferences for their outputs/services. These aspects should not be ignored while designing species specific research programs.

Research Priorities by Species

Another important issue in determining research priorities is the distribution of species specific research resources across states. In other words, if 100 rupees are available for research in the country on a particular species, how this should be allocated to different states. Bulk of the cattle research resources should be allocated to the western states of Madhya Pradesh, Rajasthan and Maharashtra, which together claim for over 30 percent of the resources.

Uttar Pradesh, Bihar, West Bengal, Orissa, Tamilnadu, Assam and Gujarat are other important candidates. It may be noted that many of these states are rainfed, where cattle is an important species to cater to the milk and draught power demand of the households. This distribution takes care of the agroecological targeting of the cattle research.

Research on buffalo should target Uttar Pradesh (21.2%), Madhya Pradesh (10%), Rajasthan (9.4%), Andhra Pradesh (9.1%), Maharashtra (8.5%) Bihar (8.3%), Punjab (5.7%) and Karnataka (5.6%). The wide dispersion of buffalo research resources indicates adaptability of buffalo to varied climatic conditions in the country. Goat research should target eastern region mainly Bihar and West Bengal. These two states together demand about 43 percent of national goat research resources. Uttar Pradesh, Rajasthan, Maharashtra and Madhya Pradesh are other important target domains.

REFERENCES

Agbamu, J.U., "Agricultural research–extension linkage systems: an international perspective", *Agricultural Research and Extension Network Paper No. 106a*, London: ODI, 2000.

Bennett, C.F., "Rationale for public funding of agricultural extension programs," *Journal of Agricultural and Food Information*, Vol. 3, 1996, pp. 3–25.

Haug, R., "Some leading issues in international agricultural extension: a literature review," *The Journal of Agricultural Education and Extension* Vol. 5, 1999, pp. 263–274.

9

Management of Agricultural Research

Management does not have a unique body of knowledge. The theories and precepts of management have been adopted from other disciplines and applied to real life situations, with a clear focus on performance of managerial roles. Management has evolved and changed considerably over a period of time. It has continuously adopted new theories and practices and replaced old ones so as to make management activity increasingly efficient.

The universal theory of management evolved in the early twentieth century has been replaced by a number of contingency theories currently in vogue. In the early twentieth century, the focus was on physical factors, viewed from industrial, engineering and economic perspectives. Subsequently, the focus shifted to productivity, with an emphasis on human factors. Managerial accounting and classical concepts of personnel and finance management were emphasized. Thereafter, many schools of thoughts evolved, each influencing the evolution of modern management. Some of these are described below.

— The *quantitative measurement school* began with concern for handling uncertainty and making decisions logically through use of mathematical models and statistical techniques. Simultaneously, there evolved the *decision theory school,* which also stresses managerial decisions, and considers the management process as a series of decisions that must be made by managers as they confront problems. Subsequently, computers used with a systems approach became the instruments of sound and logical decision making. The *systems management school* considers management as a system composed of various sub-systems (finance, accounting, production and marketing). These sub-systems are interconnected in some fashion, and operate to achieve the overall objective of the organisation.

— The *social school* views management as a system of cultural interrelationships. It deals with identifying various social groups in an organisation and integrating these groups into a complete social system. It recognises that the organisation is not isolated and must operate within the purview of social organisms in a changing environment.

— The *behaviourial science school* encompasses psychology, sociology and anthropology. It looks at management as a process of generating active interaction among individuals in an organisation to influence individual or collective behaviour. It considers human behaviour important and focal to managerial actions. This approach views various activities in relation to their impact and influence on people, who are the primary component of management.

— Currently, the emphasis is on contingency theories specific to the environmental situations in which they are applied. The *contingency management school* attempts to translate systems theory by assessing the operating factors in any situation and establishing definite patterns and relationships between those factors, which can then be used as guides in other, similar situations. Legal aspects, cultural considerations and public administration issues are also stressed.

— In the eyes of the *operational* or *management* process school, management is a unique process, which consists of certain sub-activities (planning, organising, controlling and decision making). It considers management as a series of operations and processes which provide guidelines for successful management. This is similar to the approach of the *empirical school,* which considers management from the standpoint of experience which can be generalised and certain guiding principles derived.

CONCEPT OF MANAGEMENT

Management is an activity, a process and an integrative function. It evolves as a set of effective practices by individuals in performance of specific tasks. Management is dynamic in nature and thus has many definitions depending upon the disciplinary approach adopted. An empirical definition of management would emerge in the context of specific managerial roles, which are undoubtedly influenced by a large number of organisational variables.

On this broad canvas, management can be defined as a science or art for planning, organising, implementing, controlling and monitoring, and evaluating, in order to accomplish tasks and organisational goals. These sub-activities constitute the unique management process. Management is internal to an organisation. This does not mean that it is confined to the corporate offices of organisations, but rather that it extends to research laboratories, producing centres, processing units, consuming masses and other places.

If we consider management as optimising the use of resources to accomplish goals, then the managerial functions can be easily discerned. Managers plan, organise, allocate staff, direct and control resources in an organised group effort to achieve desired objectives. This approach has many aspects, but the key variables are resources, objectives and efforts. The objectives may be organisational or individual. For example, the broad objectives of a research organisation may be to carry out innovative or adaptive research.

The objectives of individual scientists in the organisation may be to achieve professional growth and recognition, and even a sense of fulfilment in being associated with the organisation. The role of management is to organise and coordinate so as to fulfil both individual and organisational objectives in optimal manner. Human resources management involves getting the persons best suited for particular tasks, and getting the best out of them.

Thus, management is responsible for utilising the available skills to ensure the most efficient use of all human resources. Combined together, all workers should then deliver their best. Physical resources vary from organisation to organisation. In a research organisation, physical resources are laboratory equipment and apparatus, plant, machinery and facilities. The task of a research manager is to ensure the availability of technologically suitable and advanced apparatus and equipment, within the available financial resources, and to ensure that they are well maintained. In this way, scientific work should be interrupted as little as possible, if at all.

From the above, it is obvious that management is basically an exercise in doing things better. Hodge and Johnson defined management as "the process of making decisions and issuing commands on behalf of an organisation's membership groups, taking into consideration the complex of objectives, limitations and standards underlying the production and distribution of values, required to satisfy memberships' needs."

Management as a process consists of continuous decision making, necessitated by variations in goals and also the fact that lack of complete knowledge creates risk and uncertainty associated with decision making. Specific decisions are greatly influenced by the organisational goals which have to be achieved.

MANAGERIAL FUNCTIONS

Since management is coordination and integration of available resources to accomplish specific goals, it can be viewed in terms of various managerial functions: planning, organising, monitoring and controlling, and evaluating—all functions which a manager performs. These are the structural functions of a manager. In addition, a manager also directs, leads, motivates, communicates, delegates, prioritises and systematises. These are the pervasive functions of a manager.

Structural Functions

Achievement of goals requires decisions and their implementation. Changing situations give rise to new problems and hinder implementation. Problem solving necessitates decision making in various forms. One of the basic functions of a manager is to make decisions. Problems have to be foreseen, identified and available facts and data analysed under appropriate assumptions to generate alternatives. These are then evaluated in the context of the specific decision to be made and within the overall framework of the organisation.

Managers use specified decision making criteria or key variables to assess various alternatives. That is decision making. In this process, decision making ties together other functions. Participative decision making is desirable to ensure smooth implementation.

Decisions have to be implemented through action on the part of other people, and that requires planning. The plan should:

— relate to the objectives;

— fix responsibilities;

— specify procedures which must be followed;
— give a blueprint for communicating the decision to all involved and concerned; and
— have provision for participation.

Then follows implementation, and that requires organising. This involves evolving a formal structure to facilitate coordination and integration of resources for efficient accomplishment of both long- and short-term plans. Organising begins with the concept of division of work over a series of operating units, each being responsible for a particular element of implementation.

In the process of developing the total organisation structure, consideration has to be given to such issues as the degree of de-centralisation, span of control, delegation, utilisation of staff departments and chain of command. Relationships among the various operating units have to be specified. These relationships may take many forms, including authority relationships.

Implementation has to be monitored and controlled. It has four phases, namely:

— defining standards and objectives,
— determining how performance is to be measured;
— developing a reporting system, and
— taking corrective action when and where needed.

Monitoring facilitates control, to ensure that events conform to plans. Thus, in this context 'control' is concerned with progress in implementation. Through effective monitoring and control, a manager receives continuous feedback on exactly where the implementation stands at any given time with respect to achieving the objectives. If objectives are not being achieved, or if their accomplishment is behind schedule, the manager can use available information to identify the areas that are causing problems and develop alternatives to overcome these problems.

Pervasive Functions

A manager has to perform several functions which permeate any organisation. Research managers give guidance and direction to scientists under them regarding job requirements. This involves defining the work and specific role of the scientists, determining and communicating performance standards, and administering through specific policies and procedures. This is also the leadership function. After developing an appreciation of the importance of human motivation, a manager must be concerned about how the human element should be managed.

A manager should create a climate where the needs of the individual are integrated with the needs of the organisation. This means creating a climate in which the individuals can best satisfy their goals by working toward the goals of the organisation. Of particular significance in leadership is the quality of face-to-face and day-to-day interaction that managers have with their subordinates. Although many factors are involved in creating a 'result producing' climate, communication and participation are two key concerns.

One of the basic functions of the manager is to open up better communication channels. It helps the members of the organisation to develop mutual trust and understanding, and to resolve conflicts. The extent and degree to which managers work with their subordinates in a coaching and counselling capacity to help them accomplish their specific job objectives—and thus perform at their best—determines how successful they will be in their efforts at directing and leading.

Delegation of authority and responsibility is essential to make the subordinates responsible. In delegating authority, managers should see that it goes as near to the point of action as possible. Prioritising is essential in the context of limited resources and time-related targets. It facilitates systematisation of the management process.

Quality of a Manager

Since managers have multifarious activities, they have to have appropriate characteristics to support performance. Mintzberg studied various activities of managers and identified ten distinct managerial characteristics relating to interpersonal, informational and decision making qualities. Since then a host of other desirable characteristics of a good manager have been identified (Table 1).

Table 1. Characteristics of a manager

Professional	*Personal*
— Professional excellence	— Caring about people
— Analytical abilities	— Commanding respect
— Diagnostic skills	— Good judgment
— Planning skills	— Drive and energy
— Resource allocation	— Creative
— Leadership qualities	— Oriented towards excellence
— Decision making capabilities	
— Tolerance of ambiguity	
— Delegation of authority and responsibility	
— Negotiating skills	
— Coordinating and liaison	
— Human relations manager	
— Managing conflicts	

AGRICULTURAL RESEARCH MANAGEMENT

Agricultural research is the most widespread form of organised research in the world in which both developed and developing countries are engaged. As agricultural research has to be adapted to specific ecological considerations, it does not have universal applicability. The extent to which a country benefits from the research findings of other countries also depends on its own research organisations.

In the context of NARS, management broadly involves defining agricultural research goals and priorities; formulating detailed research programmes in consonance with national agricultural research strategy; assigning responsibilities to various departments; allocating financial, human and physical resources to the respective institutions; implementing approved research programmes; periodically evaluating; obtaining feedback on the impact, strengths and weakness of new technology, and incorporating them into the technology generation process; and keeping key policy-makers informed of agricultural research achievements.

Distinctive Features

Agricultural research "requires lumpy investments, involves externalities and is subject to long gestation lags". It is because of this that agricultural research has almost always to be conducted by government research organisations. In developing country situations, organisations lack vital human skills, material resources and basic infrastructure. Often, too little funds are allocated, and even they either are reduced or arrive too late. Despite all these limitations, research organisations strive their utmost to improve agricultural productivity.

Research organisations have specific, limited and clearly defined targets for their activities, which impart to them some distinctive features. Creativity is the core of a research organisation. Individuality is the key to creativity in research. Scientists and technical staff are the most important assets of a research organisation. They are the producers and creators of the organisational output.

The main outputs of research are knowledge and technology, which are too abstract to measure and evaluate on a regular basis. Because of this, very little evaluation and monitoring is done in a research organisation, particularly during intermediate stages of projects. Administrative interference and management control are minimised since scientists tend to be highly individualistic. Activities and staff are organised either according to discipline, or interdisciplinary teams are constituted, taking a project approach.

In a research organisation, the degree of command and direction is minimal, so that a subordinate has to report and be accountable to only one superior, and not to many. Even though a non-interfering environment conducive to research is created, conflicts are natural between scientists and technical personnel on the one hand, and administrative staff on the other.

Management Functions

Management of agricultural research in an institution or organisation involves four functions. The first of these functions is planning. The first step of the pianning process is to set goals. If goals are not trivial, there will be constraints to reaching them, as well as opportunities that an institution may be able to take advantage of. After identifying the constraints and opportunities, the next step is to establish objectives aimed at overcoming the constraints or taking advantage of existing opportunities.

To achieve the objectives, management identifies programmes and, within the programme limits, management and staff can determine staff, equipment, facilities and funds. The summation of all the resources needed to carry out all of an institute's planned projects is sometimes called the institute's logistics plan. Major tasks in the planning process are selecting an institute's programmes and projects. While some institutes do this on the basis of subjective judgment, many institutes follow formal models, with established selection criteria.

The second management function is organising. Once an institute plan exists, management has the basis for structuring to achieve the plan's goals and objectives, or at least to examine the existing structure to determine any changes that might be needed. There are many ways to organise or structure a research institute. A common issue is whether the institute should be centralised in one location or de-centralised in several regional centres. There are pros and cons for each option.

Another issue is whether the institute should be organised on a disciplinary or on a programme basis, again, with pros and cons for each. These approaches may be combined through a matrix structure. The third management function is monitoring and controlling. Implementation must be monitored in accordance with the plan, and any deviations identified brought back under control to keep the institute focused on achieving the plan's goals and objectives. Monitoring and controlling is an ongoing activity. To effectively monitor and control, management needs some form of management information system (MIS).

The fourth management function is evaluating. The main reason for an evaluation is to assist decision making. If no decision is to be made and no action is to be taken as a result of the evaluation, one can question 'Why spend the resources to conduct it?'. A secondary—but important—reason for conducting scheduled evaluations is for the psychological effect this has on the research management and staff. Knowing that there is to be an evaluation to measure achievement against an institute's planned goals and objectives stimulates everyone involved to make extra effort to obtain the desired project and programme results. Thus, productivity is increased.

Pervasive Functions

The functions of an agricultural research manager go

beyond the typical functions of a manager, considered to be planning, coordinating, organising, controlling and motivating. Agricultural research is a combination of three sub-systems: disciplinary areas and commodities; the management process itself; and people, who include policy-makers, scientists, research personnel, extension workers, producers, processors and consumers. The agricultural research manager has to integrate and harmonise these various sub-systems to achieve the desired goals. Thus the agricultural research manager has to:

— clearly define objectives and set priorities. The goals should be action oriented and the research should be multi-disciplinary;
— establish effective means of communication;
— synchronise with field-level requirements;
— stimulate creativity and innovation;
— evolve participatory processes; and
— manage the external interfaces with research, extension, clients and user groups, government and funding agencies.

Obviously, management of agricultural research has to be professionalised in order to perform the tasks listed above. Professionalisation is also vital, considering that research costs are spiralling, resources—particularly human—and funds are difficult to mobilise, and there is an urgent need for properly managing NARS.

ORGANISATION OF AGRICULTURAL RESEARCH

The faster the advance of basic knowledge, the greater is the productivity of applied research. Static basic research would mean diminishing returns in applied research because of increasing cost of innovation within the framework of existing knowledge. Eventually, technical change could stagnate until new knowledge emerges from basic research. Discuss the evolution of high yielding varieties (HYVs) of crop plants and subsequent stagnation in applied research. Observe that agricultural research is mostly applied.

The question arises as to the proper mix of basic and applied research that an agricultural research institute should conduct. This depends on a number of factors, such as the institute's mandate and the level of economic development of the country. Ask participants to provide examples of such factors, listing them on the chalkboard. Have participants suggest proper basic:applied research mixes for their institutions. Ask them for some arguments that would suggest that the more economically undeveloped a country is, the smaller the basic:applied research expenditure ratio tends to be.

Research may have a short- or long-term perspective. Short-term research is directed towards finding solutions to problems of immediate concern. Such problems arise in the context of the transition of agriculture from subsistence to commercial.

Long-term research is oriented towards future problems, perceived in the context of expected economic or biological developments. Long-term research is of two types:

(i) directed to problems of immediately apparent significance in the circumstances prevailing at the time the decision is made, but the execution of which is expected to be of long duration, and

(ii) research that must be initiated in the present in order to be ready for future needs based on prospects and projections pertaining to population growth, technological change and economic growth.

Technical change which permits substitution of scarce resources which are inelastic in supply by a resource like knowledge, which is less expensive and more abundant. As Arndt and Ruttan put it, the overall effect of research is to increase agricultural productivity, variously through:

— increased returns to factors of production by reduced costs or increased output,
— improved product quality,
— introduction of new production, or
— lower vulnerability through production control.

Some commonly used, but partial, productivity ratios are output per hectare, or output per unit of labour or other input. Ask participants if they can think of others. Output per unit of fertiliser applied is one. Another is output per unit of agricultural research. As agricultural scientists, we naturally hope to see an increase in agricultural output as a result of our research.

Ask participants if they, from their own experience, have any evidence of the effects of research on agricultural output. Be ready to provide your own examples. A classic example is the increase in rice yields resulting from the research by the International Rice Research Institute (IRRI) which led to the rice cultivar IR-8. Other examples are HYVs of wheat and maize. List important factors that constrain agricultural research. These would include low investment, despite the fact that the literature provides selective quantitative evidence that the internal rate of return on investments in agricultural research is high.

A 10 to 15% rate of return would be considered adequate for most investments. Observe that there are serious methodological problems in fully assessing the economic impact attributable to research. Nevertheless procedures have been developed to calculate benefits from research by:

— measuring shifts in production and supply function,
— understanding the non-conventional factors, and
— understanding how technology changes production and productivity.

The resource person should also refer to important methodological approaches such as:

— the value of inputs saved,
— the economic surplus approach (consumer-producer surplus, cost:benefit, and index numbers), and
— the econometric approach (production, profit and supply functions and their derivations).

Since research is exploration of the unknown, with a long gestation period and uncertain outputs, its financial and economic analysis both *ex ante* and *ex post* is difficult. The international funding agencies no longer undertake any such standard project analyses in relation to investment in agricultural research because of the controversy surrounding the underlying assumptions, choice of a discount rate and methodology of computation.

Nevertheless, there is ample evidence available establishing the value of returns to investment in agricultural research. Surplus cereal production in the European Union and food self-sufficiency in India are just two examples of the effects of new technologies. In both cases, agriculture was transferred from a state of serious deficit to one generating a surplus.

The first of the factors is the competing demand from other sectors, resulting from an overall lack of resources. Closely related to this is the unenlightened allocations of public funds, suggesting that policy-makers are unaware of the benefits accruing to society from agricultural research. While this is likely to be true, one might question how differently officials responsible for resource allocation would act even if they were well informed about returns to investment in agriculture.

After all, there are many factors other than economic growth that affect public decisions. You may wish to discuss some of these with participants. If we understand better the factors affecting technical change that results from agricultural research, perhaps we, as managers, will be able to maintain the current high rates of return as well as encourage greater investment in agricultural research. Under 'Market' are three issues that agricultural research managers should be aware of. The first, induced innovation, is a phrase that was coined to suggest that a country's factor endowments induce innovations, or technical change, to occur in directions which will conserve scarce factors. For example, in Japan, where labour was abundant and land scarce, innovation was induced in the direction of land-saving, biological innovations.

In the United States, where land was abundant and labour scarce, the major thrust of technical change was toward labour-saving, mechanical technology. Ask participants if they can think of examples within their own research environments where induced innovation has resulted. Economic analyses suggest that technical change may be more easily produced in some directions, e.g., labour-saving technology, than in others.

Ask participants to suggest reasons why this could be so. One might be that the fundamental research backing up technological areas may be more robust for some areas than for others. Finally, pressure groups are put forth as an issue for discussion, particularly in terms of their ability to affect the market for technical change in agriculture.

Technical change and the rate of return on investments in agriculture may be affected by the quality of management of agricultural research. Management of agricultural research includes four important management functions for agricultural research institutions. Observe that these will be discussed in the context of the organisation and management of agricultural research at various levels.

Goals for Agricultural Research

— Increasing agricultural production and productivity, to ensure food security for the rising population.

— Developing areas of untapped potential, thereby correcting emerging imbalances in growth in eastern, hilly, rain-fed and drought-prone regions.
— Meeting challenges of degradation of land and water resources, and emerging ecological imbalances, due to increased biotic pressure on land.
— Compensating for diminishing size of land holdings and fragmentation, leading to restricted management options and lower income levels.
— Addressing problems of under-employment, unemployment and malnutrition in rural areas through diversification of agriculture and promotion of horticulture, fisheries, dairy, livestock, poultry, beekeeping, sericulture, etc.
— Value addition in agriculture can only be achieved by a concerted thrust being made in increasing processing, marketing and storage facilities. These are imperative for the development of agro-processing industries, which are the key areas for development in agriculture.
— Revitalising and democratising the cooperatives for providing credit, inputs and extension support as well as enhanced marketing and processing.
— Focusing the agricultural research system to develop economically viable and location-specific technologies in rain-fed, drought-prone and irrigated areas, and strengthening institutional frameworks for farmers' education and training in improved farm techniques.
— Harnessing of scientific research, in frontier areas of science and technology, for all sections of the farming community.
— Addressing technology training and input needs of farm women, farmers living in tribal areas and other disadvantaged sections of rural society, with a view to remove the drudgery and burdens of their lives and augmenting their income.

— Accelerating the development of rain-fed and irrigated horticulture, floriculture, aromatic and medicinal plants production, and plantation crops, with full back-up support of processing and marketing, both for the domestic market and for exports.

— Encouraging efficient use of marginal lands and augmentation of biomass production through agroforestry and farm forestry.

— Increasing the utilisation of irrigation potential and promoting water conservation and its efficient management.

— Providing improved variety of seeds, agricultural implements and machinery and other critical inputs to farmers in or near their village.

— Reviving and strengthening local institutions of the farming community as legitimate instruments of decentralised planning with full participation of the local community.

— Increasing the involvement of non-governmental organisations in agricultural development and village upliftment programmes.

— Correcting the terms of trade to make them favourable for agriculture, thereby increasing the flow of resources and augmenting the rate of capital formation in agriculture substantially.

REFERENCES

Arndt, T.M., and Ruttan, V.W., *Valuing the productivity of agricultural research: problems and issues*, 1977, pp. 3-25, Minneapolis: Univ. Minnesota Press.

Ruttan, V.W., "Future research evaluation needs," in: Sundquist, W.B. (ed) *Evaluating agricultural research and productivity*, Minnesota Agricultural Experiment Station, Univ. Minnesota, Miscellaneous Publication, No. 52-1987, 1987.

Schultz, T.W., *The Economic Organization of Agricultural Research*, New York, NY: McGraw Hill, 1953.

10

Agricultural Research and Sustainable Development

Agricultural research has provided the tools and instruments that have transformed the way in which land is worked and how forest or fish resources are utilised and managed. However, the concentration of production, income and employment at regional level, in turn lead to the marginalisation of the less-favoured regions that are situated further away from processing industries and markets.

Moreover, while food security has become a less significant political issue in India since the 1980s, public health, food safety and environmental protection have gained importance. The scientific progress, economic development and agricultural policies that have contributed to successes in productivity have, at the same time, made it more difficult for many people to live from farming alone. In the light of the above, the contribution of agricultural research to rural development is being redefined to meet the challenge of a more complex countryside and of the increased services required from the agricultural sector.

Farmers now follow many different and new livelihood strategies with an increasingly diverse range of stakeholders. The crucial issue for the future of agricultural research in India, and in developing countries, is to identify the role agriculture can play in the more comprehensive process of sustainable development in rural areas and how agricultural research can support this more extensive role. Only by meeting these challenges can agricultural research make a valuable and meaningful contribution to strengthening food security and addressing the many other roles that agriculture can play in a modern society.

CHALLENGES OF AGRICULTURAL RESEARCH

Across the globe, agricultural research systems face broadly similar challenges. Countries - whether developed or developing - are affected by the globalisation of markets, economic liberalisation, advances in science and technology, natural resource management and concern for the environment. They are also influenced by new thinking about the respective roles of the public and the private sectors as well as a growing trend towards involving farmers and farmers' organisations in the process of agricultural research.

These challenges have to a great extent influenced funding and the overall orientation of agricultural research both at the national, regional and international level. Designing technologies and practices for sustainable rural development will require changes in the way research is organised, coordinated, and managed. It will affect methods of research planning and links with other institutions and sectors that have a stake in sustainable uses of land and resources. For some research projects, time frames will need to be considerably extended.

Sustainable Use of Natural Resources

The high level of intensification and specialisation of agriculture over the last decades in India has had a profound effect on rural livelihoods, biodiversity and the wider environment. The difficult challenge facing agricultural research is how it can reconcile these concerns with the need to increase agricultural productivity and ensure global food security. At present it is clear that mainstream agriculture and current land use and management policies in India do not really provide a model for a sustainable future.

Thus a key research question is what institutional and economic support system is required in order to efficiently support more sustainable land use systems and food supply chains. A major factor for future agricultural change and further development of land use systems in the region could be the expansion of non-food production and the economic return and social acceptance of set-aside areas. However a key question related to this issue is how sustainable, and, how environmentally friendly can non-food production be? A lot will depend on framework conditions and further development and implementation of good agricultural practices.

Private Research

In contrast to private research undertaken in other sectors such as electronics, information, and engineering; agricultural research carried out by private biotechnology and life science firms is controversial. Little is known about the total private spending on agricultural research and development worldwide. Private research investment in developed countries focuses on biotechnology, agri-chemicals, food processing and crop and animal technologies more suited to capital-intensive forms of agriculture with high, off-farm, value added aspects.

In developing countries, with little private sector investment, it can be expected that the private sector will play a greater role in the future when sustained profitable markets develop. However it is unlikely that private research will substantively replace public agricultural research over the next ten to twenty years.

Trends in Public Research

Public funding for agricultural research has suffered as governments in many developing countries have faced growing fiscal constraints, often as part of adjustment programmes. The reversal of funding trends, following a long period of sustained growth, has been encountered in national and international agricultural research programmes, both in rich and poor countries.

Furthermore, international donor and aid agencies no longer give agriculture, and with it agricultural research for development, the attention they once did, but provide more support to economic infrastructure, health, education and other services. There are signs of waning interest in supporting traditional agricultural research.

Mainly in developed countries, some authors question the value of investing in productivity-enhancing research in an era of chronic surpluses. Others are concerned about the environment and associate some aspects of its deterioration with activities in agriculture. Furthermore, there is increased demand by governments and donors for greater accountability and transparency in the management and conduct of agricultural research.

The publicly-funded national agricultural institutions can no longer take for granted their being the sole actor. Nor can they be assured of sustained funding without proving their relevance in the development process and demonstrating efficiency in utilising available funds. They should endeavour to forge meaningful alliances and partnerships with other stakeholders. In addition they need to broaden their research agenda to encompass a wide range of economic, environmental and socio-cultural issues besides the traditional areas of agricultural productivity, production and food safety.

Research on rural development in developing countries is only one of several instruments that can play a part in overcoming the challenges of poverty and environmental degradation. The often low priority given to rural and agricultural development in developing countries implies that even existing valuable agricultural and rural development research will not have the effect that it potentially could have.

Developing countries often adopt agricultural policies that disadvantage rural smallholders and rural entrepreneurs as against urban dwellers. The impact of joint Indian support towards agricultural research for rural development also depends on adopting the right agricultural policies in developing countries.

Indian Researchers

Past history bears evidence to the fact that Indian researchers, development agencies and governments have in many instances been forerunner innovators and global trendsetters for agricultural research and agricultural projects. For example, the drive for participatory bottom-up approaches to agricultural research and project development have been initiated and led by Indian researchers. The development and use of farming systems research approaches also received major contributions from the Indian research community.

In many fields Indian researchers are at the cutting edge of international agricultural research related to, or with relevance for, developing countries. In this context, India provides invaluable knowledge on, and models for, rural development in the developing countries. However, without funding to facilitate local adaptation of appropriate technologies such Indian capacities remain unutilised.

Contribution of Agricultural Research to Rural Development

The importance of the contribution of agricultural research is widely acknowledged but success would be determined by the ability to upscale and reorient research outputs for the benefit of the wider rural community. An increased emphasis on human resource development is crucial to meet this objective. The following are recommendations for consideration by governments, FAO, other development agencies and donors:

Advocacy for greater support

Emphasis should be placed on advocating agricultural research as an integral part of rural development among ministries of agriculture, environment and planning, regional institutions, private sector and interested donors and organisations. In this regard, the following actions are recommended:

— Bring to the attention of regional and national organisations research issues of great relevance to agriculture, agricultural research and rural development research in India and developing countries.

— Assist with a better integration of needs of India and developing countries in research programmes.

— Assist the integration of agricultural research, education and extension as well as rural development research in agriculture and poverty reduction policies and strategies.

— FAO to provide support to FAO/donor-supported regional project development for strengthening agricultural research and extension systems in India. Emphasis should be placed on integration of new technologies and their legal frameworks using participatory approaches and the improvement of information and communication systems between research and stakeholders.

— FAO to undertake a proactive role in advocating participatory research approaches in India with the involvement of a wide range of stakeholders in research priority setting, planning, monitoring and funding.

— FAO should also emphasize increased knowledge management and networking.

Focus on major issues

Greater attention should be given to issues which have not yet been adequately addressed. These include the recognised need for more transdisciplinary research, participatory approaches and the creation of information recognising and describing the complexity of rural sector development and the various roles of agriculture. The following actions are recommended:

— Encourage reorientation of research activities and create additional research capacities to study and understand the complexity of rural development processes. Such research would need to be interdisciplinary, with a stronger integration of production-oriented sciences with environmental, social and economic sciences, and an adequate and timely participation on the part of the relevant stakeholders.

— Introduce participatory and gender-sensitive approaches including social awareness raising, consensus building, and employing existing variables through mapping out diversity recognising the creativity, inventiveness and knowledge of farmers as key resources in a more integrated approach.

— Improve communication and information systems between research organisations and diverse stakeholder groups by considering the special requirements for technology and information among practitioners and the adjustment of agricultural support systems for the promotion of a more sustainable agriculture and land use.

— Examine closely the specific requirements of small countries and develop better understanding of the constraints and potentials for development in these countries in the research- extension- technology continuum.

— FAO to provide support to development of rural development indicators, monitoring and evaluation systems, impact assessment of rural development endeavours and assessment of the role and performance of civil society and NGOs in rural development.

Policy reform and priority setting

A vision of agricultural development, multidisciplinary and cross-ministerial integration and qualified human, physical and financial resources is required. Reform must lead to a better output and quality of research, and higher flexibility to respond to the evolving needs of clients. This will involve reforms in the education sector for the training of scientists, by using relevant and appropriate curricula and new ways of communication with all stakeholders involved in research development, transfer and use. The following actions are recommended:

— Build capacities for evaluating the performance of research systems to quantify and qualify the extent of reforms needed and study various reform options aiming to define more relevant research agendas and greater capacities to respond to the needs of new clients, particularly transition economies.

— Build capacities for evaluating the agricultural education and extension systems and developing new curricula, integrating the increased demands of farmers for rural non farm activities, the market economy, and new technologies and methods.

— Assist in funding for agricultural research and agricultural education reforms and for developing strategies to involve the private sector aiming to raise their contributions to reformed agricultural research and technology development.

— Donor agencies, particularly from the region, should commit to providing support to agricultural research and agricultural education reform and capacity building in India.

Strategy to maintain funding

A long-term financial commitment is necessary in order to strengthen national research capacity and infrastructure. Paradoxically, never before has the world been better placed scientifically and technologically to eradicate global poverty, hunger and human suffering. Yet it becomes increasingly difficult to do so. The actions recommended are:

— Develop a Indian strategy to prevent further deterioration in funding to agricultural research for development. The strategy should address how to raise awareness among bilateral and multilateral donors as well as among developing and transition country governments in maintaining and hopefully increasing the financial support to well-planned agricultural research for development.

Provision of Information at regional level

There is a need for information and aggregated data related to the various functions of the agriculture and gender-disaggregated data of the rural sector. In addition, in India there is often a lack of reliable data at the farm level and a deficit of the data analysis needed for research priority-setting and research planning. The following actions are recommended:

— states to collect and analyse information on pluriactivity among farm households, income combination and the various roles of agriculture and gender in the rural sector, and costs and benefits of changing land use for future rural development strategies.

— states to collect farm data on current production methods, constraints, availability and use of inputs, farm outputs, markets, etc. for problem identification and research priority setting.
— states to collect and analyse research data related to, research capacity, funding for research priority setting and identification of appropriate research methodologies and technologies.
— FAO to provide balanced information on new technologies, regulations, and encourage participation of scientists from India in relevant discussion fora and networks.

Capacity building and human resources management

There is a need to critically review agricultural education systems and to develop relevant curricula oriented towards building analytical, technical and participatory capacity and capable of addressing the agricultural and rural development needs in the sub-region. The following actions are recommended:

— Provide training for agricultural research managers and policy makers, in problem analysis, research priority setting, strategic research planning and management involving inter-disciplinary teams and participation of stakeholders to promote the sustainable use of natural resources for food security and income generation.
— Develop institutional capacity in application and regulation of agricultural biotechnologies through provision of information and training courses and address technology and the regulatory gap in India.

— Establish new ways of communication and information flow among all stakeholders involved in research development, transfer and use.

Regional and inter-regional collaboration

Sufficient resources need to be available for and targeted at such research cooperation. Efforts must be made to overcome the present weaknesses in India R&D policies towards other South Asian countries. It is recommended to:

— Create better synergies between the SAARC programmes and development aid projects, and mechanisms for setting research priorities with transition and developing countries.

— Enable increased involvement of the private sector as well as more complementarities between SAARC member states in agricultural research for development, and

— Facilitate a more concerted response by Governments, Research and Donor Organisations in support and resourcing of developing countries' efforts for improved research organisation at regional, sub-regional and national levels.

REFERENCES

Byerlee, D., G. Alex and R. G. Echeverria, "The evolution of public research systems in developing countries: facing new challenges," In: Byerlee, D and R. G. Echeverria (Eds.)., *Agricultural Research Policy in an era of privatization*, CABI, Wallingford, U.K., 2002, 300 pp.

Janssen W., T. Braunschweig, "Trends in the Organization and Financing of Agricultural Research in Developed Countries: Implications for Developing Countries," *ISNAR Research Report* 22, 2003.

Knickel, K., H. Renting and J.D. Van der Ploeg, "Multifunctionality in European agriculture," In: Brouwer, F. (Ed.), *Sustaining agriculture and the rural economy: Governance, policy and multifunctionality*, Edward Elgar Publishing Inc, 2004.

Morgounov A., L. Zuidema, "The Legacy of the Soviet Agricultural Research System for the Republics of Central Asia and the Caucasus," *ISNAR Research Report 20,* 2001.

Palmer, D., P. Munroe-Faure and F. Rembold, "Land consolidation and rural development in Central and Eastern Europe," *2nd Congress on Cadastre in the European Union,* http://www.eurocadastre.org/eng/2congress.html, 2003.

Pardey P.G., N. Beintema, "Slow magic," *Agricultural R&D a century after Mendel,* IFPRI, 2001.

Roseboom J., "Optimizing Investment in Agricultural Research, or the Quest for Prosperity," *ISNAR Research Report* 23, 2003.

11

Private Participation in Agricultural Research

In India, the land frontier is closed, irrigation is becoming more expensive, and urban growth is pulling people out of agriculture, which leaves research as the remaining major source of growth. Wealthier consumers want higher quality food and less environmental pollution, which also increases the demands on research. Since 1985, public sector investment in agricultural research in India has continued to grow, but at a slower rate.

In many of the Indian Council of Agricultural Research (ICAR) institutes, an acute shortage of operating funds has reduced scientists' productivity. In some of the state agricultural universities, the funding crunch has been even more acute. For example, the state government of Maharashtra has been gradually reducing funding every year to the state universities. In addition, the international agricultural research centers that contributed to India's growth in earlier years have had their budgets reduced considerably.

The result of the shortages of funds in some states and ICAR institutes and the weakness of public institutions for distributing public technology is that new public technology has spread very slowly to farmers., Consequently, farmers have begun planting older varieties. For example, wheat varieties in India have an average age of 9 years, versus 3 years in the United Kingdom. The private sector has held the promise of alleviating some of these problems. Private firms began funding more agricultural research in Asia and the rest of the world.

In addition, private firms have been conducting certain activities, such as commercialising and marketing new varieties more efficiently than the public sector. Thus, private research has presented an opportunity for more growth for Asian agriculture. It may have been encouraged through policy changes and public research that are more responsive to private firms' needs. Since 1985, major international trends have reshaped world agricultural input and food industries to provide more technology for developing countries through the private sector.

Barriers to international trade and foreign direct investment fell. Breakthroughs in biological sciences and favourable business conditions led to a major consolidation of biotech, seed, pesticide, veterinary, and human pharmaceutical firms into a few major life-science companies. These same firms began linking with the food industry through alliances, mergers, and acquisitions. These companies have made the latest biotechnology available to developing countries with large markets and an attractive business climate.

To fulfill the growing demand for food, fiber, and beverages, India would benefit from the private sector's playing a larger role. The international private sector appears to be ready to play a larger role, and Indian firms are increasing their investment in agricultural research. Local and international firms could fund more research and conduct research and technology transfer activities more efficiently than the public sector.

AGRICULTURAL DEVELOPMENT

The major impetus to Indian agriculture was given during the late 1960s and early 1970s, with the advent of the so-called "green revolution technology." Agricultural production in India has rapidly grown since then. The index of agricultural production rose from about 86 in 1970-71 to 176 in 1996-97. Since the last major drought in 1987-88, agricultural growth has been good, despite marginal setbacks in 1991-92 and 1995-96.

Given the limits to area expansion, the increased production has essentially been the result of rising land productivity. These higher yields in turn were achieved by the use of modern agricultural inputs, including irrigation, chemical fertilisers, as well as improved and high-yielding seeds and pesticides. About 80 million hectares of cropped area were irrigated in the 1990s. While the area under irrigation consistently increased over the years, only about 38 percent of the gross cropped area had access to irrigation in the 1990s.

The dependence of Indian agriculture on rainfall continues to be significant. The use of high-yielding varieties (HYVs) rapidly grew during the 1970s and 1980s; the rates of growth of HYV use seem to have declined in the 1990s. Along with the use of HYV seeds, the production and distribution of certified and quality seeds and the consumption of chemical fertilisers has also increased. The index of fertiliser consumption rose from about 40 in 1970-71 to 259 in 1996-97. Fertiliser consumption per hectare rose from about 13 kilograms to 77 kilograms.

About 56,000 tons of pesticides (technical grade material) was consumed in 1996-97. This is a marked increase from 24,000 tons in 1970-71. The 1990s experienced some deceleration in the quantity of pesticides used; the index declined from 167 in 1990-91 to 125 in 1996-97. These indices are based on official statistics. Estimates provided by industry sources suggest an increase in the consumption of agro-chemicals even during the 1990s.

The pesticide market in India was dominated by insecticides (76 percent); the share of herbicides (13 percent) and fungicides (11 percent) in the agro-chemical market was rather small. Tractor production in India has been rising since the early 1970s. More than 191,300 tractors were manufactured in 1995-96, while the reported production in 1990-91 was only about 138,500. No reliable estimates are available for production of diesel engines and electric motors for irrigation.

According to industry sources, approximately 500,000 to 600,000 diesel engines were produced for agriculture, and the demand for such engines rose by 5 percent per year during the 1980s and early 1990s. Recent changes in India's cropping patterns and trends in capital formation also need to be highlighted. Changes in cropping patterns can contribute to increases in agricultural yields per hectare if the area shifts from low-yielding to high-yielding crops. Such changes also have implications for the demand patterns of agricultural inputs as the use intensity of these inputs varies significantly across crops.

The rate and nature of capital formation in agriculture also impinges on the rate of agricultural growth. The share of food grains in gross cropped area declined from about 75 percent in 1971-72 to about 67 percent in 1994-95. Within the food grains category, the percentage of area under coarse cereals (maize, sorghum, and millets) declined from 28 percent to 17 percent. The share of area under pulses also marginally declined. While rice retained its share of 23 percent, the share of wheat increased from 11 percent to 14 percent.

Oilseeds were the major gainers among nonfood grains; their share in cropped area increased from 9 percent to about 15 percent. Cotton, which suffered a bit in the 1980s, improved its share in the 1990s. Crop pattern shifts in favour of superior cereals and nonfood grain crops, such as oilseeds, can ceteris paribus increase the demand for agricultural inputs as these crops consume relatively more inputs per unit of land.

Despite the decent performance of Indian agriculture, which augurs well for the agriculture-related industry, a few disconcerting aspects need to be emphasized:

— The annual rate of growth of agricultural production was lower in the 1990s, than in the late 1980s; and

— The annual rate of growth of input use (high-yielding varieties, fertilizer, irrigation, and power) was also lower in the 1990s than in the late 1980s.

This downturn is partly due to the relatively slow growth of real plan expenditure on agriculture since the early 1990s. Estimates suggest that, after peaking in 1978-79, gross capital formation in agriculture declined afterwards. India's agricultural economics improved again in the 1990s, with investment going up to Re 70 billion in 1996-97.

The stagnation of the 1980s, which continued into the early 1990s, was essentially due to the decline in public sector gross capital formation. Private investment did not rise fast enough to compensate for the relative decline in state-sponsored investment. The 1990s saw a reversal of this trend with private investment rapidly rising. In 1996-97, as much as 84 percent of the total investment came from the private sector.

One can view these trends in two ways. It can be argued that public investment is gradually being replaced by private investment, and one need not be concerned about the decline in the state's role in agriculture-related investment. The other view could be that given the complementarities between public and private investments, private investment would have risen even faster if the public investments had continued to grow rapidly.

In fact, the decline in the rate of growth of fertiliser use, irrigation, and high-yielding varieties could have been arrested, given the complementarities of use in these inputs and the importance of state support in expanding the use of high-yielding varieties, irrigation, and fertiliser. Limited availability of concessional agricultural credit could have also contributed to this process. Overall, the picture that emerges is that agriculture's growth since the beginning of the 1990s has been good enough to support the growth of agriculturerelated businesses. However, there are some indications that the growth performance could have been better if public investment in agriculture had not declined in recent years.

PRIVATE R&D EXPENDITURES

Most agricultural research conducted in India is very applied. The types of research conducted by private firms, the amount of expenditure. Estimates of levels and growth of private research and development (R&D) expenditures by different industries are presented in table 1.

Table 1. Research expenditures by private firms and state-owned enterprises, India, 1984-95

Industry	Research expenditures (1995 dollars) 1984-85	1994-95	SOE in 1994-95	Increase
	Millions		Percent	
Seeds	1.33	4.93	0	271
Agricultural machinery	3.70	6.48	13	75
Fertilisers	6.80	6.65	67	-2
Pesticides	9.00	17.02	15	89
Veterinary pharmaceuticals	.90	2.72	5	203
Sugar industry	.90	2.49	1	177
Food processing	1.27	10.33	1	712
Vegetable oil processing	.07	.14	0	99
Total	23.97	50.75	16	112
Public research	206.22	347.90		69

These estimates are supplemented by some firm-level estimates for different agricultural subsectors from the

Centre for Monitoring Indian Economy. In the 1990s, the private sector spent between $39 and $43 million on food and agricultural research in India. In industries where interview data were insufficient, industry-level data for 1994-95 from DST were used. The largest research expenditure was by the food industry, followed by pesticides; the seed industry and agricultural machinery were both $5 million to $6 million. The poultry, fertiliser, and feed industries made smaller investments in R&D.

Private research in India has grown rapidly since 1985. Between 1985 and 1995, private research expenditures at least doubled. This is faster than public agricultural R&D, which grew 69 percent. The DST data underestimates growth—particularly in industries with many new entrants—because it takes a number of years for new firms to get approval.

In addition, some firms do not get approval because there is little benefit from this designation—some tax reductions and a substantial cost in paperwork. In the Pray and Kelley survey of the Indian seed industry, firms not approved by DST conducted 24 percent of the research of the firms surveyed. Similarly, the DST estimates of private R&D for firms producing chemical fertilisers and tractors are much lower than the estimates derived from CMIE data: $2.2 million versus $7.6 million for fertilisers and $5.6 million versus $19.9 3 The data compiled by these sources are not strictly comparable.

For example, the estimates of R&D expenditures compiled by the DST are based on the data made available by firms having DSTrecognised R&D units in different industry groups. Not all recognised firms provide this information every year and not all firms doing R&D are recognised by DST. Consequently, these data usually underestimate the R&D in the sector.

The CMIE estimates are based on data compiled from company annual reports. All companies listed in the

Mumbai Stock Exchange are covered. Here again, not all firms' annual reports are available to CMIE every year. At times, firms do not report R&D expenditures in their annual reports. By and large, for the organised sector, CMIE estimates are more robust than DST estimates. million for tractor firms. The higher estimates from CMIE are largely due to CMIE's including all R&D, chemicals, fertilisers, trucks, and machines of tractor firms.

The most rapid growth took place in food processing, followed by the seed industry, pharmaceuticals, and the sugar industry. Pesticides research almost doubled during 1985-95. Actually, pesticide research probably grew even more rapidly, but several firms that increased their research, such as Monsanto and DuPont, were not registered as research companies with DST. Tractor research also substantially increased: it was 75 percent higher in 1995 than in 1985, then declined the next 2 years. The industry in which R&D declined, according to DST data, was fertilisers.

Seed Firms

India has a large number of seed firms, but only a few have large operations. About half of seed sales are by public corporations. Since the mid-1980s, large Indian firms and multinationals have entered the Indian seed industry. According to Pray and Kelley, firms with some foreign ownership in 1995 accounted for about one-third of the private half of the seed market, and large Indian firms accounted for 23 percent.

Most seed firms conduct breeding research to develop new hybrids based on inbred lines that have been developed in the public sector, international agricultural research centers, or parent companies. A few large programs conduct research to develop their own inbred lines. At least three seed companies have major biotechnology labs in India conducting basic biological research. One lab spent about $700,000 in 1997, had 11 Ph.Ds out of a total of 34 scientists, and had collaborative research with some top university biotech labs in the United States.

Two of these labs work on hybrid rice issues—understanding hybridisation or identifying markers that can screen for grain quality. These labs are also transforming cotton and vegetables with bacillus thuringiensis and other genes. A number of companies test transgenic varieties developed either from their own programs or from foreign programs. The Department of Biotechnology, which must approve any field trials of genetically modified organisms, reports that there have been 28 field trials of transgenic crops since 1996, of which 95 percent were by private firms. These trials have been for cotton, mustard, tomato, eggplant, and cabbage. Soybean trials had been approved but were not yet in the field, and trials for potatoes and tobacco were in progress.

Animal Feed Industry

There are very few corporate players in the animal feed industry; the bulk of animal feed is producted in the small or cooperative sector. Besides, many farmers prepare their own animal feed. The share of the corporate sector in the animal feed industry is near 33 percent and has been rising. This market has about 40 relatively large firms; the others are small.

The total estimated market size was Re 23 billion in 1995-96. We have R&D data for only three major firms. Godrej Agrovet and Hindustan Lever emerged as two significant corporate players in this market. Hindustan Lever increased its market share by acquiring some firms and expanding capacities.

While the R&D to sales ratio for Hindustan Lever declined somewhat during the 1990s, the R&D intensity increased for the other listed companies. Unlike the other two firms, Hindustan Lever is a large diversified firm of which their R&D estimates include expenditures on activities other than animal feed. The research arms of animal feed firms essentially test new ingredients, study ways to reduce anti-nutritional factors, and test new additives provided by other firms.

Sales of Flowers

This sector attracted a lot of investment in the 1990s, with multinational corporations' (MNC) establishing export-oriented units. There is no estimate of the total sales of flowers. About 15 major firms sell Re 150 million worth of flowers per year. Estimates of R&D expenditures (many of these companies are bio-tech companies) are difficult to obtain. The focus of R&D in this area was in testing foreign varieties and developing management techniques to grow flowers efficiently.

Agricultural Chemicals and Crop Protection

Several large firms operate in this segment, and the extent of rivalry is high. Multinational corporations also figure significantly in this sector. Many international mergers and acquisitions have impinged on the market structure of this industry. Hoechst and Schering became Agrevo in 1994. Ciba-Geigy and Sandoz agricultural chemicals merged to become Novartis in 1996. In December 1999, Hoechst and Rhone-Poulenc merged to create Aventis. In the 1990s, between 70 and 80 firms engaged in producing agricultural chemicals. The top 10 firms had a 63-percent share of a Re 31 billion market.

Only a few small firms produce active ingredients of pesticides. However, some small firms formulate the final pesticide composed of the active ingredients and inert chemicals. Research by the crop protection industry was also almost entirely applied. The two main research activities of these industries are conducting efficacy tests on chemicals new to India and developing new methods of producing commercial chemicals. The first type of research is conducted primarily by the subsidiaries of foreign firms because they are the source of almost all new pesticides. They test the chemicals that have been commercialised elsewhere to find out how effective they are against Indian pests and diseases under Indian climatic conditions, application methods, and market conditions.

These tests are required by the companies to ensure that product meets their specifications and the registration requirements of the government. In addition, the chemicals must be tested for their effect on the environment, workers' health, and animals. Foreign and local firms spend some research resources trying to develop the most effective package of practices for the use of these chemicals. Some of these packages probably qualify as integrated pest management.

Table 2. Seed imports, India, 1988-95

Year	*Cereals (maize, sorghum, millets)*	*Pulses*	*Oilseeds (primarily sunflower)*	*Vegetable seed*	*Total*
			Tons		
1988-89	0.64		0.11	11.34	14.14
1989-90	.13	.02	.14	82.52	82.81
1990-91	.80		5.09	77.59	83.50
1991-92	3.37		373.66	51.33	428.39
1992-93	1.73	.05	22.50	121.31	148.08
1993-94	.76		58.32	170.02	235.06
1994-95	2.19	.01	33.46	414.34	459.91

The main research activity of local firms has been in developing new methods of producing the active ingredient of pesticides discovered elsewhere. This allows the local firm to produce chemicals originally produced by a method kept secret by the inventor or protected by process patents. Local and foreign firms test different formulations for their products and different combinations of their products and other chemicals that might complement them. A few local firms are starting to develop research programs to develop new active ingredients for pesticides using standard chemical synthesis methods.

A larger number of local firms and at least one foreign firm were considering natural products to use as pesticides. Local firms seemed to be concentrating on neem tree extracts, plant growth regulators, and a few other things traditionally used in Indian agriculture. One foreign firm has a program to actively collect plants that might have biological activity. They then screen these plants and send a handful of the most promising ones to Europe each year. In 20 years, this program still has not led to a new commercial product.

Manufacturing Tractors

About 12 firms were manufacturing tractors in the 1990s in India. Mahindra and Mahindra, TAFE, and Punjab Tractors gained in market share, while Eicher, Escorts, and HMT lost. However, all major players, except HMT, increased their R&D intensity in the 1990s. The decline in the R&D expenditures at HMT tractors is understandable since it was for sale.

The tractor industry underwent major restructuring in the 1990s. While the demand for tractors has grown consistently over the years, its rate of growth declined in 1997-98, as compared with 1995-96. It is expected to decline further. Small tractors were exempted from excise taxes until 1994. Inputs used to produce these tractors were also excluded from duties.

The 1994-95 budget made the final product excise-free, and the companies had to pay duties on raw materials. Subsequent changes in the value-added tax meant that tax advantages of making small tractors declined further. To add to the problems of small tractor manufacturers, in 1995-96 the government also extended subsidies on tractors to the high horsepower versions. This was an extremely important step as tractor demand is significantly influenced by availability of subsidies and soft loans.

Larger tractors are not only more efficient in the field but are also more useful for transporting products. Over the years, the inadequacy of transportation infrastructure has resulted in the use of tractors to transport men and produce. Given these policy and market changes, it is expected that firms will try to upgrade the horsepower range of their tractor production. Tractor firms conduct a substantial amount of research in India. A major thrust has been to develop higher horsepower tractors that are also affordable to Indian farmers.

For Indian firms, this means developing tractors with higher horsepower. For example, Eicher, known for its low-horsepower tractors, started selling a 38-horsepower model and hopes to produce 42- and 62-horsepower tractors in the future. A large part of R&D by almost all tractor firms is spent to gear up for the production of larger tractors. In anticipation of a boom in demand, the market players had enlarged their capacities. This expansion, along with the entry of new players, resulted in underused capacities.

Exports are seen as a source of improving rates of capacity use. But such a strategy also requires capability to produce large tractors, as the external markets do not prefer small machines. The export markets, especially the United States and Europe, also have certain design and quality specifications that are different from those in the Indian market. Firms are also conducting R&D to conform to these standards to enlarge their export markets.

Most leading players in the market tried to obtain technology for large tractors through multinational participation. In addition, the three new entrants—New

Holland, Sami-Greaves, and John Deere-L&T—are entering with more advanced foreign tractor models. For foreign firms, this means modifying large tractor models developed for the United States, Europe, and South America to be less expensive, yet efficient, and safe on the road; one of the main uses of Indian tractors is hauling crops to market.

Diesel Engines

There are about 31 manufacturers of diesel engines in the corporate sector; not all of them, however, make engines for the agricultural sector. Slow, low-horsepower (<10 horsepower) diesel engines are reserved for small firms. As a result, the bulk of the agricultural demand is satisfied by about 800 small manufacturing units spread over the country. The small sector produces about 500,000 small diesel engines every year, mainly for irrigation but also for sugarcane crushers and generating power.

The corporate sector contributes another 90,000 engines for irrigation. Estimates of R&D are unavailable for this segment. The diesel engines produced by small firms are based on outdated Petter and Lister models. Concessional credit and government subsidy has been restricted to slow, low-horsepower diesel engines. As such financial support drives demand to a significant extent, policy has contributed to technological obsolescence.

Small industry reservation and financial support for slow, low-horsepower engines has meant that producers did not spend on R&D to upgrade the old models. The diesel engine story is, therefore, somewhat similar to the tractor story in which government-support for small tractors helped their persistence in the Indian market. However, unlike the tractor industry, not many new entrants have started producing new engines with multinational technologies. Field Marshall in Rajkot is one exception, which is trying to introduce HATZ diesel engines through German collaboration.

Marine Products

A significant number of large firms are engaged in the production of marine products. But their share of the estimated 1997 sales of Re 123 billion was only 10 percent. Most of these major players are engaged in exporting marine products. The relatively high R&D expenditure of ITC, Ltd., in this segment is misleading, because ITC is a large conglomerate firm, and separate estimates for R&D in the marine products sector are unavailable.

Similar data are unavailable for other firms as well. Our discussions with some large firms in this sector revealed that whatever limited R&D is conducted is to meet the quality standards for exports, especially to the United States and Europe. These efforts have intensified with the European Union's banning some Indian exports. Most Indian firms export unprocessed marine products. Some firms are trying to move up the value chain and are undertaking research for this purpose. Only a few firms are trying to enter the ready-to-cook market, with research to develop such products.

Poultry Industry

India has several poultry-breeding firms—more than any country outside the United States and Europe. These firms use pure lines from the United States or European firms and breed them in Indian conditions. Therefore, the chickens must survive extreme heat and some cold, because few barns have climate controls, and they must tolerate less hygienic conditions.

The firms also have to be competitive in the Indian market structure in which the commercial hatcheries are separate firms from the suppliers of grandparent stock. Therefore, the chickens must lay a large number of eggs. The poultry industry consists largely of small firms. No estimate of the total market size is available. Including

Venkateswara, the largest group, there are approximately 10 major corporate firms in the poultry market. The total sales of chickens for these firms was Re 1.6 billion in 1993-94 and 1995-96. R&D data are available for only three firms.

Since Venkateshwara Research, Venkateshwara Breeding Farm, and Western Hatcheries actually belong to the Venkateshwara group, effectively we have data for only one company. We were informed that other than the Venkateshwara group, only a few other firms undertake any significant R&D in the poultry sector. While the market share of Venkateshwara in processed chicken declined a bit in the 1990s, their R&D expenditures significantly increased. In fact, the data from the firm show an even higher level and increase in R&D expenditures for the group, from Re 52 million in 1993-94 to Re 129 million in 1997-98.

Food Processing Industries

Data on the structure of the food processing industry are difficult to compile as this sector includes a large variety of products. Therefore, only flour milling is included here. Food processing industries do a limited amount of agricultural research to improve their inputs. For example, beer companies try to improve the quality of the grain they use for malting, and tobacco firms try to reduce the cost of the tobacco they buy while retaining a certain quality standard. Pepsi has identified and popularised superior tomato varieties for the Punjab.

Most research by the food industry is, however, concentrated on developing new products and manufacturing processes. There are about 30 relatively large firms engaged in the manufacture of flour milling products. Their share in the market was, however, very small, only 10 percent of the Re 58 billion, annual market. Very few of them report R&D expenditures.

The entry of firms in flour milling, especially wheat flour, is a recent phenomenon. By and large, the packed and branded wheat flour could not withstand competition from the small producers. Consequently, the R&D activity initiated in the early 1990s has not increased. Some respondents indicated that increasing the shelf life was the major focus of this research. Apparently, some research is also being conducted to retain the softness of the kneaded wheat flour for relatively long periods.

Technology Transfer

Even in agriculture, where new technology is often embodied in plants and animals that are very sensitive to changes in climatic, soil, and pest conditions, some technology can be transferred with very little adaptive research. Some of this technology comes in as finished or almost finished inputs and the quantities can be indicated by input imports. For other technology, the knowledge is purchased and the product is made in India.

Finally, some technology is brought in as a part of direct foreign investment by foreign firms. Imports of agricultural inputs are very limited for India. For example, seed imports are negligible except in vegetables. Sunflower was the only field crop with appreciable imports of commercial seed for only 1 year, 1991-92. Even that was only 7 percent of the total commercial use. The volume of imports of vegetable seed, for which restrictions on trade were eliminated except for a small tariff, increased much more than field crops but were still small.

Imports of many other inputs, such as tractors and diesel engines, were not permitted. Pesticide inputs increased. Imports of technology through multinational firms can be indicated by proposals approved by the Indian Government. Of the 8,795 approved proposals for which we have data, 1,582 (18 percent) were in the agriculturerelated sectors. These approved proposals anticipated equity flows of $31 billion, 12 percent of which were to flow into agriculture-related sectors.

The proposed participation of multinational corporations in the agricultural business industry was mainly in the form of equity flows and establishing export-oriented units. Licensing of technology was the third most important MNC linkage. Financial and technical participation of MNCs in these industry groups is likely to enhance technology flows as well. Food processing of various kinds (instant semiprocessed foods, meat preparations, and other food products), as well as vegetables, fruits, and flowers are the main sectors attracting MNC participation.

Interestingly, input industries, such as fertilisers, pesticides, and agricultural machinery, have not attracted many projects. However, the input industry is not as diversified as the other product groups, and the number of firms in the input industries is also small in relative terms. Consequently, the entry of even a few MNCs may have significantly increased the competitive pressures in the input segments. That may have been the effect of MNC entry on some food processing segments as well.

Effect of Private Research

In the debates about intellectual property rights and biotechnology, critics of the private sector continually argue that private firms will drive up prices of inputs and not provide farmers any benefit from research. In contrast, most economists argue that although the price of improved inputs, such as hybrid maize, may increase, farmers' total costs of production will decrease because they need less of other inputs.

The reduction in needed inputs can be measured as partial factor productivity, such as output per hectare or total factor productivity. If output per hectare increases, less land will be needed to produce the same amount of output.

Thus, a farmer is saving on his costs of land by using the new technology. If total factor productivity increases, farmers' costs are reduced by using the new technology.

Measuring the effect of private research is beyond the scope of this study. However, three types of evidence indicate that private research has increased productivity and thus reduced farmers' costs of production. First, evidence is available from the companies interviewed about the effect of their R&D effort on partial productivity measures. Second, three studies measured the effect of private research on output per hectare and total factor productivity. Third, studies of industrial research and technology purchase in India show a positive effect of R&D and technology purchase on total factor productivity of industry. This suggests that research on new processes by the food industry and input industries increases productivity, which will eventually benefit farmers and consumers.

Table 3. Increases in poultry efficiency due to poultry research, India, 1981 and 1996

Item	*Unit*	*1981*	*1996*
Broilers:			
Days to 1.5 kg body weight	Number	47	38
Feed conversion	Percent	2.5	1.85
Mortality	Percent	3	2
Layers:			
Eggs production to 72 weeks	Number	270	315
Feed efficiency	Percent	145	134
Mortality (72 weeks)	Percent	8	6

The industries interviewed provided several examples of productivity increases due to their research. One example is from Venkateshwara Hatcheries (VH): improved VH breeds increased the productivity of their layers and broilers considerably through breeding. Table 3 shows that the number of days required to rear broilers to a marketable-size bird was reduced by 20 percent.

The amount of required feed declined by 26 percent, and mortality also dropped. Table 3 shows that the number of eggs from their layers increased by 17 percent, while the feed requirement declined by 7 percent and mortality declined. These data come from VH poultry operations; data were available from other commercial farms. It seems likely that the productivity increases on other commercial farms would be less.

Table 4. Effect of private and public hybrids on yields, India, 1998

Crop/State	*Private*	*High-yielding variety*	*Estimation technique*
Sorghum, Andhra Pradesh	0.0027	-0.09	Random effects
	(1.92)	(1.54)	
Sorghum, Karnataka	.00833	.44	Random effects
	2.34)	(2.99)	
Sorghum, Maharashtra	.008	.23	Fixed effects
	(1.54)	(1.88)	
Pearl millet, Andhra Pradesh	.0007	-.084	Fixed effects
	(.27)	(1.1)	
Pearl millet, Karnataka	-.0002	.39	Random effects
	(.11)	(3.2)	
Pearl millet, Maharashtra	.01	.02	Fixed effects
	(1.91)	(.32)	
Maize, Andhra Pradesh	.023	-.11	Fixed effects
	(2.27)	(.7)	
Maize, Karnataka	.005	.77	Random effects
	(.48)	(1.7)	
Maize, Maharashtra	.04	.13	Fixed effects
	(3.33)	(.96)	

Another success of the VH group was in producing vaccines that are less expensive than some commercial vaccines and more reliable than government vaccines. They also developed vaccines that provide protection against diseases for which no other vaccine exists. In fact, they developed one vaccine that no other country has developed. This vaccine is for a form of hepatitis that has become a serious problem for poultry in India since 1993. The commercial sale of this vaccine in recent years has greatly reduced deaths from this disease and increased industry productivity.

Several recent studies measured the effect of technology developed and introduced by the private sector. A study of maize used farm-level data from six states in 1994-95 to show that the adoption of hybrid maize led to yield increases of about 1 ton per hectare over improved open-pollinated varieties. In total, this led to an increase in maize production of 1.1 million tons. To obtain these increases, farmers had to increase fertiliser, irrigation, and pesticide use in addition to adopting hybrids. Therefore, increased output is not entirely productivity growth. Since most hybrids in 1995 were from private firms, most of this gain was due to private research.

A study by Ramaswami, Pray, and Kelley looked at the factors that influence the partial productivity index, yield per unit of land. The dependent variables were cotton, maize, sunflower, sorghum, and pearl millet yields. The independent variables included a measure of the spread of high-yielding varieties (HYVs), the spread of private varieties, the proportion of irrigated crop area, fertiliser use, the number of regulated markets, and the length of roads in the district, in addition to profitability of the crop, a trend variable, and variables measuring rainfall.

The basic model is augmented by interaction variables of HYVs with private varieties, irrigation, and fertiliser use. Since private varieties have been significant in these crops only recently, their analysis was confined to the period since 1985. Private hybrids' effects on yields are positive and statistically significant in five of the nine crops and provinces and close to significant in a sixth case. Table 4 summarises the results of the regressions. These estimates provide the first econometric evidence that private plant-breeding affects crop yields in developing countries.

This is particularly impressive because the region examined is in the semi-arid tropics where private research is not expected to have much effect. The only study that had considered how the benefits of private hybrids were divided was conducted by Pray et al. It examined the increases in seed prices and increases in farmers' yields of hybrid sorghum and pearl millet in Maharashtra and Gujarat. For hybrid sorghum, at most 18.5 percent of the benefits were captured by the seed companies through higher prices, while 81.5 percent went to farmers as the value of increased production minus the increased cost of seed. For hybrid pearl millet, only about 6 percent of benefits were captured by seed firms.

More than 90 percent of the benefits from private pearl millet research went to farmers. Using this same data, Ribeiro estimated the social rate of return to private plant-breeding research in India to be 38 percent or more. A study of total factor productivity of crop production, by district in 13 major states of India from the 1950s to the 1980s, also provides evidence of the effect of private research and technology transfer. The study found that private research and technology transfer, advances in agricultural research outside India, and public research all made major positive contributions to total factor productivity growth in the crop sector.

The social rate of return from investments in private research was very high—exceeding 100 percent—which suggests that most benefits from private research go to farmers and consumers rather than input companies and that society believes that there is a substantial underinvestment in private research. Studies of the experience of Indian industry found that technology imports reduced local R&D by a small amount but increased the productivity of Indian firms.

The lost productivity from the small decline in research was more than offset by increased productivity from the imported technology. Indian firms did not need to use their own resources to reinvent technology developed elsewhere and could concentrate their research instead on new products and processes that could not be purchased from abroad.

TECHNOLOGICAL EFFECT IN PIPELINE

Perhaps the most important technology in the pipeline is hybrid rice, which the public and private sectors are racing to commercialise. Sixteen private seed firms reported that they are breeding hybrid rice, and several of these breeding programs are quite substantial. MAHYCO and SPIC have biotech research programs on hybrid rice. Farmers grew hybrid rice in 1997-98.

Most of the 250,000 hectares under hybrids in 1997-98 was under private hybrids, with Proagro, Pioneer, and MAHYCO leading the way. These private hybrids are based on public lines from the Indian Council of Agricultural Research (ICAR), International Rice Research Institute, and China, but they are private hybrids. Yields are often 1 ton per hectare more than the best conventional varieties, and yields of hybrid seed are high enough to make the 1-ton increase commercially viable.

The main problem is the grain quality of hybrid, which is low. Thus, most private firms and many public research institutes are concentrating on improving grain quality. Another important technology that several private firms and a number of public institutes are researching is single-cross hybrids of maize. Companies reported that singlecross hybrids produce 10 to 30 percent higher yields than double-cross hybrids in trials in India.

Seed of single-cross hybrids is still not being marketed in significant quantities because of the high cost of seed production and the ease with which it can be copied by contract farmers and competitors. But this technology will be supplied as intellectual property rights are strengthened. Among the first biotechnology products likely to be approved are *Bacillus thuringiensis* (Bt) cotton, which allow farmers to reduce the number of insecticide applications from 15 or more to 3 and to achieve higher yields. Another likely early approval of a genetically engineered crop is hybrid rapeseed that yields 10 to 20 percent more than improved local varieties. Other near approvals are pest-resistant tomatoes, cabbage, and eggplant.

A new generation of pesticides were being introduced in 1998 that are effective against some pests that have grown resistant to older pesticides and are much safer for people and the environment. Several new wheat herbicides were approved for control of Emperata grass, which is resistant to the herbicides on the market in India. Companies estimate that use of these herbicides, which are much more environmentally friendly, will increase yields by 20 percent. These new products will be more expensive.

An International Maize and Wheat Organisation survey found that the ratio of the price of hybrid maize seed to the price of commercial grain in selected developing countries in 1990 ranged from 1.3 in China to 25 in Cameroon, with India at 4.2. Single-cross hybrid seed in the United States and Europe cost more than 30 times the cost of grain. Thus, it is probable that as Indian farmers adopt better double-cross and single-cross hybrids, prices will rise. However, as of 1998, prices have been held down by intense competition among private seed companies, public seed firms, and farmers who save seed.

There is no evidence that this will change soon. Even if the firms in which Monsanto has some ownership—MAHYCO, Cargill, and E.I.D-Parry—were to merge, they would have less than 14 percent of the commercial seed market, which is a small part of the total seed planted that primarily comes from farmers. No one in 1998 knew the price of pest-resistant hybrid cotton with Bt in it. However, the experience of China gives some indication.

In China, the price of Bt cottonseed (variety not hybrid) increased from yuan 5 per kilogram to yuan 42 per kilogram. However, because of the higher quality of the seed, the quantity sown could be reduced to one-quarter of the amount of traditional seed. Thus, the seed cost per unit of land doubled, rather than increased five times. In return, farmers saved 10 to 20 pesticide applications, saving money for chemicals plus the cost of labour to apply the chemicals. Bt cotton is very popular in Hebei Province where it was released.

INCREASE OF R&D INVESTMENTS

Liberalisation in India and changes in multinational firms' strategies are major causes of the increase in research and technology transfer to India. An analysis of seed industry data indicates that local and foreign companies increased their research in response to liberalisation. In 1998, foreign firms, such as Monsanto and DuPont, invested in new agricultural research stations, and John Deere entered the Indian market for the first time with its latest line of tractors.

The Foreign Investment Promotion Board data on multinational corporations' proposals for entry into different industry groups also suggest such a trend. According to neoclassical economic theory, firms seek to maximise expected profits. The expected profits to a firm from investing in research are a function of the expected benefits and costs of research and development of a commercial product discounted by an interest rate.

The expected benefits will be based on the expected size of the market, the share of the market the firm hopes to capture, and the expected price of the new product. Firms will calculate the expected market size based on current market size and growth rates for this industry. They will estimate their expected share of the market by looking at their current market share in the industry, the strength of intellectual property rights in the country, and technical means of protecting their product from copying.

The expected price will be based on current prices of similar products plus their ability to keep other firms from copying the product and competing against them. Economists use the term "appropriability" to describe a firm's ability to capture economic gains from research.

The expected costs of research depend on the availability of needed technology elsewhere in the world. The environmental specificity of foreign technology will determine whether there are opportunities for adaptive research or direct material transfer. The availability of technology from public institutes, which can be adapted or modified through local research, can reduce research costs. The salaries and benefits of scientists, engineers, and technicians are important components of research costs, as are laboratories, experiment stations, and the supplies to run them.

Market Size and Growth

Increased agricultural research was partially a function of increased demand for agricultural products and modern agricultural inputs. The size of Indian markets for agricultural inputs has grown substantially since 1980. The private sector supplies most of the equipment for minor irrigation, half of the certified and quality seed, half of the fertiliser, most of the pesticides, and most of the tractors.

In the 1990s, rapid growth continued in tractors, power tillers, and minor irrigation. Production of seeds and pesticides, likewise, continued to grow, but that growth was in a different dimension than is captured by the measures used earlier. In the seed industry, more expensive private hybrids replaced subsidised public hybrids and public varieties. In the pesticide industry, newer, more expensive pesticides which require only 40 to 50 grams of technical material per hectare replaced older, less expensive pesticides that required 2,000 grams per hectare.

The result was more sales measured in value terms. India is one of the largest national markets for agricultural inputs in the world. It ranks first in the number of tractors produced and sold. It is also one of the largest fertiliser and pesticide producers and consumers. Part of the increase in research in the 1990s was due to foreign input firms' deciding that the Indian market was simply too large to ignore, even if many policies were not conducive to high profits. Like the input industries, the Indian food industry is one of the largest in the world, and it is rapidly growing.

India is expected to become the world's most populated country early in the 21st century. Demand for processed foods, poultry products, dairy, and meat has also rapidly grown, increasing demand for improved livestock technology. Output by the food industry doubled between 1980-81 and 1995-96. Production of poultry products more than doubled in the 1980s and increased another 25 percent from 1990-91 to 1995-96. Milk production increased by 70 percent in the 1980s and 22 percent from 1990-91 to 1995-96.

Production of livestock feed also rapidly grew from 1.2 million cubic meters in 1980-81 to 2.9 million cubic meters in 1995-96. These increases were driven largely by increased demand created by gains in per capita income. In addition to increased market size, the market share of large private Indian and foreign firms has increased since 1980. Very large Indian firms and firms with foreign ownership of more than 40 percent were excluded from the seed and biotechnology industries until 1986. Half of the active ingredients of pesticides had to be formulated by small firms, and all of the agricultural implements industry was reserved for the small firms until the 1990s.

The elimination of those restrictive policies allowed large firms into the seed market and permitted the manufacturers of the active ingredients of pesticides to increase their market. It is also expected to induce tractor manufacturers into the implements industry. The market shares of government-owned corporations have declined in the seed industry and in tractors as government sales grew more slowly than private sales.

In addition, some public companies in this sector, like HMT tractors, have been included in the state's disinvestment program. Both factors increased the market share of private firms. Changes in the input markets in the United States and Europe have made the markets of countries of Asia, in general, and India, in particular, very attractive, relative to their traditional markets. From World War II to the late 1970s were boom years for agricultural input firms in the United States, Europe, and, to a lesser extent, Latin America.

With the stagnant or declining growth of the 1980s, most U.S. companies reacted by reducing costs. By the early 1990s, having squeezed costs as much as they could, many of them started to look to developing countries, Eastern Europe, and the former Soviet Union for further growth. John Deere, DuPont, and Monsanto in the early 1990s expanded into developing countries, including India.

In addition, due in part to developments in biotechnology, agricultural chemical and pharmaceutical companies shed their traditional chemical business and bought biotech and seed firms to transform into life-science companies. These companies invest large sums of money in basic research to develop new drugs, seeds, and agricultural chemicals, which they then try to sell worldwide to pay for their research.

Appropriability

Appropriability—the ability of a firm that owns new technology to capture some benefits that users of the technology obtain—can be due to several causes. First, laws like patent acts can give owners temporary monopolies, enabling them to raise prices and profit from selling the technology. Second, the structure of the industry may allow firms to capture some benefits.

Monopoly or oligopoly power in a market can give inventors high enough prices to profit from technology. Third, the technology may allow firms to keep others from copying a technology thus giving inventors market power. Fourth, firms can simply keep inventing and stay ahead of their competition. This also would allow them to charge more.

Since 1985, there have been only a few changes in appropriability in India. The laws and enforcement of intellectual property rights have not changed since 1972, when new chemicals, pharmaceuticals, and food and agricultural products were excluded from product patent protection. However, as earlier mentioned, the markets have become more competitive.

India signed the Uruguay Round of GATT (General Agreement on Tariffs and Trade, now the World Trade Organisation) and was committed to a *sui generis* plant breeders' rights (PBR) law and strong process patents on biotechnology products by January 1, 2000. The country also plans to issue product patents for new chemicals, pharmaceuticals, and food and agricultural products by 2005. PBR legislation and amendments to the patent act were proposed and debated by several different Indian Administrations but not passed.

In addition, India must protect trade secrets and extend liability to third parties that induce breach of a trade secret, and protect test data which is submitted for obtaining marketing approval of a new product. India's signature to the GATT agreement may have raised the hopes of research-based firms for stronger intellectual property rights, but not too much. In the seed sector, appropriability increased through technical means.

Hybrid seed is becoming viable in additional crops. Developments in hybrid rice seed production after 1985 led to the commercial adoption of hybrid rice in 1997. In addition, several systems for producing hybrid rapeseed seem possible. These developments led to private investment in a number of new research programs on rice and rapeseed, increasing total seed research.

Regulatory changes allowed foreign firms to increase their share of ownership in all sectors. This enables the foreign owner to appropriate a larger share of profits from new technology back to the firm's headquarters where much of the research is conducted. The potential for enhanced profitability will increase the interest of foreign firms to invest in research in India. The increased entry of foreign firms and some large Indian firms into agricultural input and agricultural processing industries has increased the competitive pressure on all firms in these industries. Firms have to innovate more rapidly to keep their market share. They try to appropriate the gains from their research by staying ahead of the competition.

Cost of Innovation and Probability of Success

Firms must weigh the expected benefits, which are based on market size and appropriability, against the cost of innovation and the possibility that the innovation will fail to generate the expected sales. The cost of innovation and probability of success are a function of the state of basic science, quantity and price of scientific inputs such as scientists and labs, and agroclimatic differences between the place for which a new product was designed and India.

Advances in basic science can lead to new possible products from applied research. One major breakthrough since 1980 has been in biotechnology. Profits from plant biotechnology products are no longer a dream, but rather a reality in the United States, Canada, and Argentina. This has drawn a number of seed firms and agricultural chemical firms to invest in biotechnology research in India.

In 1985, the first survey found that Hindustan Lever and a few other firms had started to work on plant biotechnology. Now Hindustan Lever, Tata Tea, and at least three seed firms have substantial plant biotech labs in India. The output of more applied public research can also stimulate private research. During 1985-98, public research institutes in India and international centers provided considerable stimulus to private plant-breeding research. Participants of a study of the Indian seed industry reported that the International Center for Research in the Semi-Arid Tropics (ICRISAT) was a very important source of germplasm by 65 and 80 percent of the sorghum and millet breeding firms, respectively.

The Indian Council of Agricultural Research (ICAR) and State Agricultural Universities (SAU) were very important for 66 percent of the cotton breeders. Sunflower was the only major crop for the international centers and ICAR/SAU were reported not to be important sources; joint venture partners are the most important sources of sunflower breeding material. Singh, Pal, and Morris documented the importance of ICAR and CIMMYT germplasm as the basis of private maize research. The downsizing of ICRISAT and SAU and weak funding of ICAR meant that many well-trained and experienced scientists were available to private firms to lead and staff their administrative and research positions.

The negative side of weak funding is that less public science and technology is available from these institutions in the long run. Some effect of declining funds can be seen at ICRISAT, which has stopped having the sorghum and pearl millet field days, at which ICRISAT scientists displayed and distributed samples of their latest hybrids, varieties, and inbred lines. Another way to reduce the cost of research is to learn from ideas and innovations elsewhere in the world. Since 1985, several types of innovations have become more accessible to private firms.

The reforms of the seed industry in the late 1980s made inbred lines and earlier generation germplasm more easily accessible. The reforms made it easier to import varieties for research and finished varieties for a few years to try them on a commercial basis. The admitting of foreignowned firms meant that they brought in germplasm and new ideas that spilled over to local firms. Some of our survey respondents expressed the fear that new intellectual property rights (IPRs) might result in less sharing of germplasm and other research materials.

Regulatory reforms that reduced the time for new chemicals to be approved influenced pesticide companies to bring in more products, which stimulates local research on these products. Research by the agricultural chemical industry depends primarily on the number of new chemicals introduced. Each new chemical needs a minimum amount of research to ensure that it works well in India and that it is registered.

REFERENCES

Desai, B.M., "Budget: A Retrograde for Agriculture," *Vikalpa*. Vol. 22, No. 2, April-June 1997, Pp. 23-26.

Pray, Carl E., Sharmila Ribeiro, Rolf A.E. Mueller, and P. Parthasarathy Rao, "Private Research and Public Benefit: The Private Seed Industry for Sorghum and Pearl Millet in India," *Research Policy*, Vol. 20, 1991, pp. 315-324.

Unni, K.K., "Role of Agro-chemical Industry," *Agricultural Development Paradigm for the Ninth Plan Under New Economic Environment*, B.M. Desai (ed.), Oxford and IBH Publishing Co., Pvt., Ltd., New Delhi. 1997.

Bibliography

Alston, J.M., *et al.*, "A Meta Analysis of Rates of Return to Agricultural R&D: Expede Herculern", *Research Report* 113. IFPRI, Washington, D.C, 2000.

Birthal, P. S., P. K. Joshi and Anjani Kumar, "Assessment of Research Priorities for Livestock Sector in India", Policy Paper 15, *National Centre for Agricultural Economics and Policy Research*, New Delhi, 2002.

Byerlee D. and M. Morris, "Have we underinvested in research for marginal environments?" The example of wheat breeding in developing countries, *Food Policy*, 18: 381-393, 1993.

Chandrashekhara, K. and K.N. Ganeshaiah, "AICRP programmes: In the need to rebirth", *Current Science*, 85 (2), 25 July, 2003.

Evenson, R. E., C. Pray and M. W. Rosegrant, *Agricultural research and productivity growth in India*, Research report 109, Washington,DC: IFPRI, 1998.

Falck-Zepeda, J. B., G. Traxler and R. G. Nelson., "Cotton GMO adoption and Private Profitability," In: *Genetically Modified Organisms in Agriculture- Economics and Politics*, G. Nelson (ed.), London: Academic Press, 2001.

Fan, S., P. Hazell, and S. Thorat, *Linkages between Government Spending, Growth, and Poverty in Rural India:* Research Report No. 110. IFPRI, Washington, 1999.

Fujisaka, S., L. Harrington and P. Hobbs, "Rice-wheat in South Asia: systems and long-term priorities established through diagnostic research", *Agricultural Systems* 40. 169-187, 1994.

Heston, A. and R. Summers., "The Penn World Table (Mark 5): An Expanded Set of International Comparisons, 1950-1988," *Quarterly Journal of Economics*, May 1991, pp.327-368.

Pray, Carl E., Sharmila Ribeiro, Rolf A.E. Mueller, and P. Parthasarathy Rao, "Private Research and Public Benefit: The Private Seed Industry for Sorghum and Pearl Millet in India," *Research Policy*, Vol. 20, 1991, pp. 315-324.

Ruttan, V.W., "Future research evaluation needs," in: Sundquist, W.B. (ed) *Evaluating agricultural research and productivity*, Minnesota Agricultural Experiment Station, Univ. Minnesota, Miscellaneous Publication, No. 52-1987, 1987.

Schultz, T.W., *The Economic Organization of Agricultural Research*, New York, NY: McGraw Hill, 1953.

Unni, K.K., "Role of Agro-chemical Industry," *Agricultural Development Paradigm for the Ninth Plan Under New Economic Environment*, B.M. Desai (ed.), Oxford and IBH Publishing Co., Pvt., Ltd., New Delhi. 1997.

Index